D1293589

INTERMEDIATE ANALYSIS

A MATHEMATICS TEXTBOOK
UNDER THE EDITORSHIP OF
Carl B. Allendoerfer

M. S. Ramanujan

DEPARTMENT OF MATHEMATICS
UNIVERSITY OF MICHIGAN

Edward S. Thomas

DEPARTMENT OF MATHEMATICS
STATE UNIVERSITY OF NEW YORK, ALBANY

INTERMEDIATE ANALYSIS

THE MACMILLAN COMPANY
COLLIER-MACMILLAN LIMITED, LONDON

© Copyright, The Macmillan Company, 1970

All rights reserved. No part of this book may be reproduced or transmitted in any form or by any means, electronic or mechanical, including photocopying, recording or by any information storage and retrieval system, without permission in writing from the Publisher.

First Printing

Library of Congress catalog card number: 73–90876

The Macmillan Company
Collier-Macmillan Canada, Ltd., Toronto, Ontario

Printed in the United States of America

517
R165i

122296

Preface

This book began as a set of notes for a post-calculus course taken by prospective mathematics majors at the University of Michigan. The main objective of this course, and hence that of the book, is to provide the student with a feel for the way in which contemporary mathematicians design and build the machinery they use.

This approach dictates the initial subject matter of the book, set theory.

Every student of mathematics soon becomes aware of the fact that any mathematical concept can be stated ultimately in terms of sets, set operations, functions, and relations. Thus a knowledge of set theory is essential for the mathematician. Besides this, elementary set theory as presented here is simple enough, yet abstract enough, to provide the beginner with an abundance of problems on which to cut his teeth.

The more complicated set theoretic notions we have explored are used in that part of the book devoted to intermediate analysis. This portion includes a discussion of what the real number is, the elementary topology of the real line, and two chapters devoted to infinite series.

The reader who works through the text will have seen the development of mathematical notions from the extremely simple through the rather difficult.

We have adopted and stressed the viewpoint that rigor is essential in mathematics at each step in the development of an idea. That is, the idea must be rigorously defined and its consequences verified in every detail. On the other hand, once this point has been driven home, we expect the reader to supply many of the details himself. The majority of the exercises are designed to give the reader practice in supplying proofs and many of them fill gaps in the text.

Although the construction of the real numbers and verification of their properties are relegated to the appendixes, the reader is expected to use his intuitive knowledge of the real numbers throughout the text as a source of examples.

We have assumed, in several places, a knowledge of differential calculus—particularly in the chapters on infinite series. In addition to this there are several places in the text where the principle of induction is used. Finally, some exercises involving the concept of finiteness precede the definition of this term; these may be omitted without impairing the flow of the text.

We are indebted to several of our colleagues who read and commented on various portions of the manuscript and to many of the students who were subjected to preliminary versions of this book and who pointed out errors and omissions.

We acknowledge with thanks the aid we have received from the Department of Mathematics of the University of Michigan and, most especially, from the secretarial staff who typed much of the manuscript.

Finally, special thanks are due our wives for their encouragement during the writing and rewriting of this book.

M. S. R.
E. S. T.

Contents

CHAPTER 0

Some Remarks About Logic

1. INTRODUCTION

This chapter is a brief, and by no means thorough, introduction to some of the rules by which the deductive process in mathematics is carried out. Almost everyone is aware of these rules, perhaps implicitly, but experience indicates that they may be forgotten or not believed just when needed most.

It is hoped that the remarks which follow will serve as a set of guidelines, not obstacles. The reader may have an easier time with this chapter if he realizes that it has little or no mathematical content of the type he recognizes. Rather it is metamathematical. We are merely going to make some remarks and observations about mathematics and, more specifically, about what a mathematical assertion is and what constitutes a proof. We also introduce some commonly used terminology and notation, much of which may already be part of the student's vocabulary.

2. ASSERTIONS

Here are three mathematical statements which we number for future reference.

2.1. If x and y are positive real numbers, then xy is positive.

2.2. If f is a differentiable function, then f is continuous.

2.3. If f is a continuous function, then f is differentiable.

Each of these statements has essentially the same form. We have one or more objects, denoted by letters (variables), under consideration and two statements about these objects, the hypothesis and the conclusion. Each statement is of the form "If (hypothesis), then (conclusion)." In statement 2.1, for example, the objects are two numbers, denoted by x and y. The hypothesis is that both numbers are positive and the conclusion is that their product is positive.

Each statement is an assertion to the effect that a certain hypothesis implies a certain conclusion in the sense that if any object (or set of objects) satisfies the condition stated as the hypothesis, then that object (or set of objects) also satisfies the condition stated as the conclusion. In this sense each of the "if, then" statements can be rewritten in the form "(hypothesis) implies (conclusion)." We call such statements *assertions* or *implications*.

A mathematical theory consists in large part of a body of valid assertions which are derived, via standard rules of logic, from basic definitions and axioms. Assertions which are considered especially important and/or "pretty" are given special titles, for example, *theorem*, *corollary*, *proposition*. Obviously, it is important to know what it means for an assertion to be valid (true) or invalid (false). We now discuss these concepts and touch on the related idea of a proof.

3. VALIDITY AND SOME REMARKS ON PROOFS

Let H and C be statements concerning one or more objects. We shall say that the assertion "H implies C" is *valid* or *true* provided that if an object or set of objects satisfies H, it also satisfies C; otherwise, we say the assertion is *invalid* or *false*.

Let us consider assertions 2.2 and 2.3 in light of the above definition. As everyone knows, assertion 2.2 is valid; your favorite calculus book contains a proof of it. It follows that since the function $f(x) = \sin x$ is differentiable, it is continuous. Similarly, $f(x) = e^x$ is differentiable, hence continuous. Indeed, once we know that 2.2 is valid, we can keep getting specific examples of continuous functions as long as our stock of differentiable ones holds out. One common mistake people make is to assume that the reverse of this process is true, that is, that an implication is valid if it is valid in a number of specific cases. The problem, of course, is that some cases not tested may show the assertion to be false. In connection with assertion 2.2, we also observe that the implication says nothing about the continuity of a function which is not differentiable; there are nondifferentiable functions which are continuous and ones which are not continuous.

Turning now to assertion 2.3, let us note that the functions $f(x) = \sin x$ and $f(x) = e^x$ satisfy both the hypothesis and the conclusion. By looking at these two functions, one might jump to the conclusion that assertion 2.3 is valid. However, it is not valid, because the absolute value function $f(x) = |x|$ is continuous, but, at the origin, it fails to be differentiable.

With these illustrations in mind, we list some observations about validity and proofs of assertions.

3.1. *To prove that "H implies C" is valid it does not suffice to exhibit examples of objects satisfying H which also satisfy C.*

3.2. *To prove that "H implies C" is invalid, it suffices to exhibit an object which satisfies H but not C.* This process is called finding a *counterexample*; thus $f(x) = |x|$ is a counterexample to assertion 2.3.

3.3a. *If "H implies C" is valid and an object does not satisfy C, then neither does it satisfy H.*

3.3b. Equally important, *to show that "H implies C" is valid, it suffices to show that if any object does not satisfy C, then neither does it satisfy H.*

Statement 3.3a is just a matter of juggling words in the definition of validity. Statement 3.3b can be illustrated by the following example. Suppose we wanted to prove assertion 2.2, "If f is a differentiable function, then f is continuous." We could go about this directly by showing that the condition for continuity (the δ-ε condition) follows from the condition for differentiability (the difference quotient condition). This would be what is usually called a *direct proof*.

There is another way to prove this assertion: We could show that if a function is not continuous, then it is not differentiable. Conceivably this would be done by writing down in terms of δ's and ε's what it means for f not to be continuous and, using this, to show that some difference quotient does not behave as it must in order for f to be differentiable. Proofs employing this tactic are often called *indirect proofs* and are quite common (although in the example chosen it so happens that a direct proof is neater).

There is a further refinement of the indirect proof, called *proof by contradiction*. The idea is this: To prove that "H implies C" we assume that there is an object which satisfies H but not C and, by some sort of argument, arrive at a conclusion which contradicts a known fact. The usual proof of the statement "If $x = \sqrt{2}$, then x is irrational," is an example of such a proof. One assumes that $\sqrt{2}$ is rational (that is, $\sqrt{2} = m/n$, where m and n are integers) and, by some tricky manipulations, shows that this contradicts some known algebraic facts.

In the above we have shown one way of not proving an assertion, given descriptions of the main varieties of proofs and discussed the idea of a counter example. We have not touched on the really crucial questions of what a proof is and how a person thinks one up. It requires a highly formalized language to give an accurate definition of the word "proof." We probably all agree that, intuitively, a proof of the validity of the implication "H implies C" is a logical chain of reasoning which starts with the statement that some object satisfies H and, using facts already proved or axioms, ends with the statement that the object also satisfies C. Strictly speaking, this is pretty much mumbo jumbo, since we do not know what a "logical chain of reasoning" is. Happily, it turns out that we can get by without a formal definition, because people involved in mathematics over a period of time tend to get by some sort of osmosis a feeling for what constitutes a mathematical proof.

Probably one learns to recognize whether an assertion has been proved or not by attempting to construct his own proofs and by criticizing those of others. This brings us to the second question. Proofs are concocted of a combination of experience, intuition, insight, ingenuity, and, sometimes, good luck. This book is designed to help the reader acquire some measure of the first three of these commodities; the last two are in great demand and each person must supply his own.

We close this section with a warning. It is easy to write down assertions which are meaningless, in the sense that the hypothesis or the conclusion contains variables which are not well-enough modified. The following example illustrates this: "If x is a real number, then $x + y = 10$."

The objects under consideration appear to be real numbers, but there is a variable, y, in the conclusion which is not quantified (modified) in the hypothesis. We have no way of proving or disproving this assertion without additional information about y.

For example, if we add to the hypothesis the requirement that $y = 5$ we get: "If x is a real number and $y = 5$, then $x + y = 10$." This is false, since $x = 2$ is a counterexample. But if we add the requirement that y denotes the quantity $10 - x$, we get a valid assertion: "If x is a real number and $y = 10 - x$, then $x + y = 10$."

The reader should assiduously avoid meaningless statements of the sort just described, although "dangling variables" may become harder to detect, as the level of abstraction increases.

4. NOTATION AND TERMINOLOGY

Instead of writing "If H, then C" or "H implies C," we frequently use the symbolism "$H \Rightarrow C$." Thus assertion 2.1 becomes: "x and y are positive real numbers $\Rightarrow xy$ is positive."

Consider now two statements, which we shall denote P and Q instead of H and C. The *converse* of the implication "$P \Rightarrow Q$" is the implication "$Q \Rightarrow P$"; thus the converse is obtained by reversing the roles of hypothesis and conclusion. Note that assertion 2.2 is the converse of assertion 2.3 and, of course, vice versa.

Obviously an assertion may be valid and its converse invalid. If, however, it happens that both "$P \Rightarrow Q$" and "$Q \Rightarrow P$" are valid, we say that P and Q are *equivalent* and write "$P \Leftrightarrow Q$." This last symbolism is frequently read "P if and only if Q"; here the "only if" corresponds to the arrow $\Rightarrow$ and the "if" to the arrow $\Leftarrow$.

For example, everyone knows that if x is a positive real number, then so is $x/10$ and, conversely, if $x/10$ is a positive real number, so is x. Thus we have "x is a positive real number if and only if $x/10$ is a positive real number."

Besides forming the converse, we can alter the assertion "$P \Rightarrow Q$" in other ways. One such way is to form the assertion "not $Q \Rightarrow$ not P", where "not P" and "not Q" are the negations, or denials, of the statements P and Q. The statement "not $Q \Rightarrow$ not P" is called the *contrapositive* of "$P \Rightarrow Q$." The contrapositive of assertion 2.2 is the assertion "If a function f is not continuous, then it is not differentiable."

The point we want to make is that *a given assertion and its contrapositive are equivalent* in the sense *that they are either both valid or both invalid.* In particular, to prove "$P \Rightarrow Q$" it suffices to prove "not $Q \Rightarrow$ not P." This is precisely the content of observation 3.3b and our remarks about indirect proofs in Section 3.

We close this chapter by establishing the following convention. *From now on* when we mean to say that the implication "$P \Rightarrow Q$"

is true we shall merely write "$P \Rightarrow Q$." This keeps us from writing "is true" every time we state a theorem. If we wish to express the fact that "$P \Rightarrow Q$" is false, we write "$P \not\Rightarrow Q$" and say "P does not imply Q."

EXERCISES

Choose a mathematics book which has proofs in it. Practice writing some theorems, propositions, and so on, as formal implications. Form converses and contrapositives. Classify the proofs, if possible, à la Section 3.

CHAPTER 1

Set Theory

1. INTRODUCTION

A strong argument can be made that vagueness and outright ambiguity are useful, perhaps even essential, features of ordinary languages. By contrast, one of the features which characterize mathematics and make it useful is the precision required of its statements and arguments. Experience indicates that such precision is best achieved by using a special language. This chapter is a brief introduction to that language and, indeed, to a special framework, called set theory, in which mathematics is done.

The first order of business is to learn enough set theory to communicate ideas mathematically. The notation and terminology which we introduce can best be mastered by doing exercises. We have included drill problems, designed mainly to fix the basic concepts through repetition, and problems which involve translating statements into the language of set theory. There are also a number of useful little facts which seem to occur often enough in proofs to warrant inclusion as exercises but not as theorems.

2. BASIC TERMINOLOGY

Intuitively, a set is a collection of objects which possess one or more common properties and which are called the members or elements of the set. Pretty clearly, this is not a good definition of the word "set" since it involves the word "collection," and there is no more reason to know what a "collection" is than to know what a "set" is. A little reflection indicates that we cannot define every term we use; it is necessary to start somewhere with terms which are not defined.

In set theory we take as undefined the word "set" and the notion of being an element of a set. Customarily, capital letters are used to denote sets and lowercase letters to denote elements of sets.

Our first bit of notation is a shorthand device to indicate membership in a set. If S denotes some set and x some object, then we shall express the fact that the object denoted by x is a member of the set denoted by S simply by writing $x \in S$.

For variety, the words "set," "collection," and "family" will be used interchangeably, as will the expressions "is a member of," "is an element of," "belongs to," and "is in."

3. SPECIFYING SETS

There are two methods commonly used to specify what the members of a set are. The simplest and most obvious of these is to list those objects, and only those, which are members of the set. The list is inclosed in braces to indicate that it is the set of these objects, and not the individual objects, which is under consideration. Thus the set whose members are the first four positive integers may be denoted by $\{1, 4, 3, 2\}$, by $\{4, 1, 2, 3\}$, or even by $\{1, 1, 2, 4, 4, 2, 3, 3\}$.

Unfortunately, the process of compiling lists becomes tiresome and, indeed, physically impossible for most of the sets we wish to consider. To alleviate this problem we most often define a set by telling precisely what conditions an object must satisfy to be a member of the set. The conditions may be expressed in ordinary language, or in set theoretic language, or as a combination of both. Obviously, then, a given set may be described in many ways.

The set whose elements are the integers 1 and 2 is specified by any of the following: "the set whose members are the first two positive integers"; "the set of all solutions of the equation $x^2 - 3x + 2 = 0$"; "the set of positive integers x such that $-2 < x \leq \frac{7}{3}$."

The following notation is frequently useful when sets are defined in the above way. For the moment, let $P(x)$ stand for one or more statements about elements in a set A. Here the letter x is merely used to denote a typical element of A. The set of elements for which $P(x)$ is true is denoted by

$$\{x \in A \mid P(x)\}.$$

This may be read "The set of x in A such that $P(x)$."

For example, if we use I to denote the set of integers, then the set consisting of the first two positive integers may be described in this brace notation by $\{x \in I \mid x^2 - 3x + 2 = 0\}$ and by $\{x \in I \mid x \text{ is positive and } -2 < x \leq \frac{7}{3}\}$.

4. RELATIONS BETWEEN SETS

Let A and B denote sets. We say that A is a *subset* of B, or is *contained in* B, or that B *contains* A, and we write $A \subseteq B$ or $B \supseteq A$ in case every element of A is an element of B.

Thus, if I denotes the integers, then

$$\{x \in I \mid x^2 - 4 = 0\} \subseteq \{x \in I \mid x \text{ is even}\}$$

and the set $\{1, 3, 5\}$ is a subset of the set of all odd integers.

The sets A and B are said to be *equal* and we write $A = B$ provided that $A \subseteq B$ and $B \subseteq A$. This definition will be appreciated more after several proofs involving the equality of sets have been completed. Notice that the definition is merely a way of saying that A and B have exactly the same elements.

In case A is a subset of B but A is not equal to B, we say that A is a *proper* subset of B and write $A \subset B$.

Negation of any of the relationships $\in$, $\subseteq$, $=$, or $\subset$ is denoted by a slash mark. Thus $x \notin A$, $A \nsubseteq B$, and $A \neq B$ may be read, respectively, "x is not an element of A," "A is not contained in B," and "A is not equal to B."

At this point we include for future reference a seemingly trivial observation. Suppose A and B denote sets. The statement that A is not contained in B is equivalent to the statement that there is at least one element of A which is not an element of B; in particular, *if no such element of A exists, then A is contained in B.*

EXERCISES

In each of the following problems you are given a six-letter alphabet. In the first two problems the letters denote sets and in the third they denote statements about sets. For the moment, define a *word* to mean a list using some or all of the letters such that if one letter, say X, follows another, Y, then the set (statement) denoted by Y, is a proper subset of (implies but is not equivalent to) the set (statement) denoted by X. For instance, if X denotes the set $\{1, 2, 3, 4\}$ and Y the set $\{2, 4\}$, then YX is a word but XY is not.

1. Here x denotes an integer.

$$A = \{x \mid x^2 = |x|\},$$
$$B = \{x \mid 2x^2 - x - 1 = 0\},$$
$$C = \{x \mid x \text{ is even}\},$$
$$D = \{x \mid |x| \leq \tfrac{3}{2}\},$$
$$E = \{x \mid \cos x\pi/2 = 0\},$$
$$F = \{x \mid \text{every integer is the product of } x \text{ and some integer}\}.$$

(a) Show that no word contains the same letter twice.
(b) Find all the three-letter words you can.
(c) Find all the four-letter words you can.
(d) Find all the five-letter words you can.
(e) Either find a six-letter word or show that there are none.

2. Here f denotes a function defined on the real numbers with real values. This is essentially a problem in rephrasing some results of calculus in terms of sets.

$$A = \{f \mid f \text{ is differentiable everywhere}\},$$
$$B = \{f \mid f(x) = 3x^2 - 1\},$$
$$C = \{f \mid \text{for every } x, f(x) \neq 0\},$$
$$D = \{f \mid f \text{ is continuous everywhere}\},$$
$E =$ the set consisting of the functions $f(x) = e^x$ and $f(x) = |x| + 1$,
$F =$ the set of all polynomial functions.

(a) Find the letters corresponding to sets which are not subsets of any other set.
(b) Find all three- and four-letter words.

3. In this problem S and T are sets.

A: S is a subset of T but T is not a subset of S.
B: S contains some element which is not in T.
C: T is not equal to S.
D: Some member of T does not belong to S.
E: One of S, T is not contained in the other.
F: Neither of S, T contains the other.

(a) Rewrite each statement employing, as best you can, the symbols $\subset$, $\subseteq$, $=$, and their negations.
(b) Write down all words.
(c) Which statements (if any) are equivalent to A? To B? To C? To D? To E? To F?

5. OPERATIONS ON SETS

In this section we introduce four basic ways of combining sets to get new sets. Three of these are quite natural constructions.

In what follows A and B denote sets.

The *union* of A and B, denoted $A \cup B$, is the set of all objects which are members of A or members of B.

The *intersection* of A and B, denoted $A \cap B$, is the set of all objects which are members of both A and B.

The *complement* of B in A, denoted $A - B$, is the set of elements of A which are not members of B; we sometimes read $A - B$ simply as "A minus B."

Finally, the *Cartesian product* of A and B, written $A \times B$, is the set whose elements are all the *ordered pairs* (a, b), where $a \in A$ and $b \in B$. We emphasize that the ordering is important here. An ordered pair (a, b) has a first coordinate, a, and a second coordinate, b, and if an ordered pair belongs to the Cartesian product $A \times B$, then its first coordinate must be an element of A and its second an element of B. Two ordered pairs (a, b) and (a', b') in $A \times B$ are *equal* if and only if $a = a'$ and $b = b'$.

To illustrate the above remarks, let A consist of the integers 1 and 2 and let B consist of the letters x and y. Then $A \times B$ has four elements: $(1, x)$, $(1, y)$, $(2, x)$, $(2, y)$ and $B \times A$ has four: $(x, 1)$, $(x, 2)$, $(y, 1)$, $(y, 2)$, but $A \times B$ and $B \times A$ have no elements in common.

We pause briefly to introduce one additional concept. It was stated at the beginning of this section that the operations to be introduced would yield new sets from old ones; in particular, the intersection of *any* two sets is supposed to be a set. It follows that there must be sets having no elements.

Accordingly, we now postulate the existence of a set having no elements. Suppose A is such a set and B is any set. From the remark at the end of Section 4 it follows that $A \subseteq B$. Thus if A and B are sets having no elements, then each is contained in the other and hence, by definition, they are equal.

To summarize, there is one and only one set which has no elements and it is a subset of every set. This set is called the *empty set* or the *null set* and denoted by $\varnothing$.

6. VENN DIAGRAMS

It is customary in mathematics to use pictures to illustrate the situations one is considering. Examples include the ruler and compass constructions of geometry and the graphing one does in calculus. These sketches never prove anything by themselves, but they do serve as valuable aids to the intuition and sometimes provide the insight needed to solve problems.

In set theory we use what are called Venn diagrams to illustrate problems involving unions, intersections, and complements of sets. Sets are represented as regions enclosed by curves. If several sets are being considered, the corresponding regions may be marked by shading, crosshatching, or even by coloring. Combinations of these sets can then be visualized easily.

We illustrate with some examples.

Example 1. Suppose A and B are the sets corresponding to the regions marked with vertical and horozontal lines, as shown in Figure 1.1. Then $A - B$ is the region marked with vertical lines

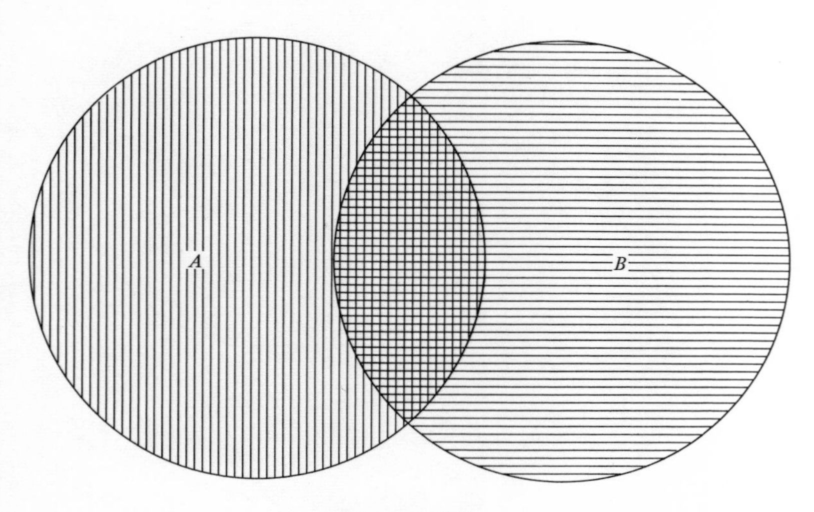

FIGURE 1.1

only and $A \cap B$ is marked with both vertical and horizontal lines (Figure 1.2).

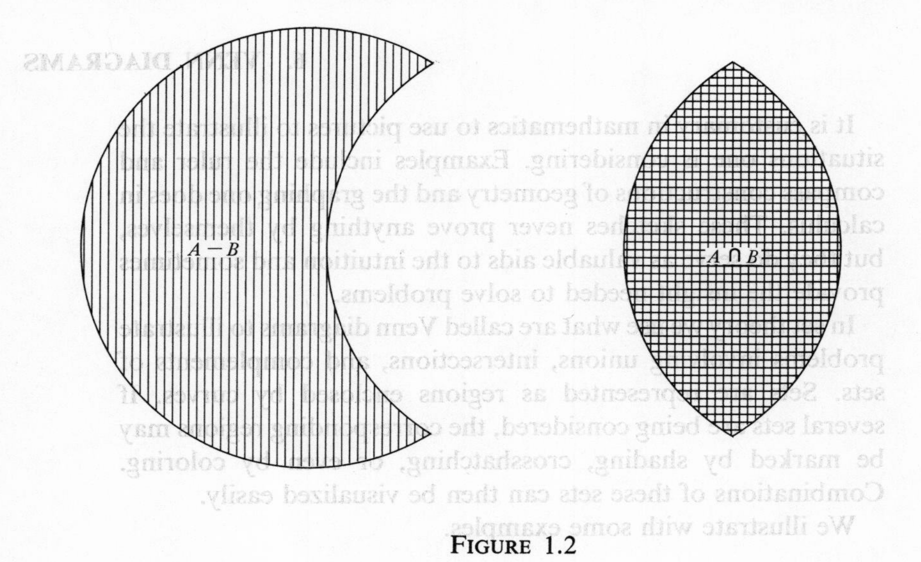

FIGURE 1.2

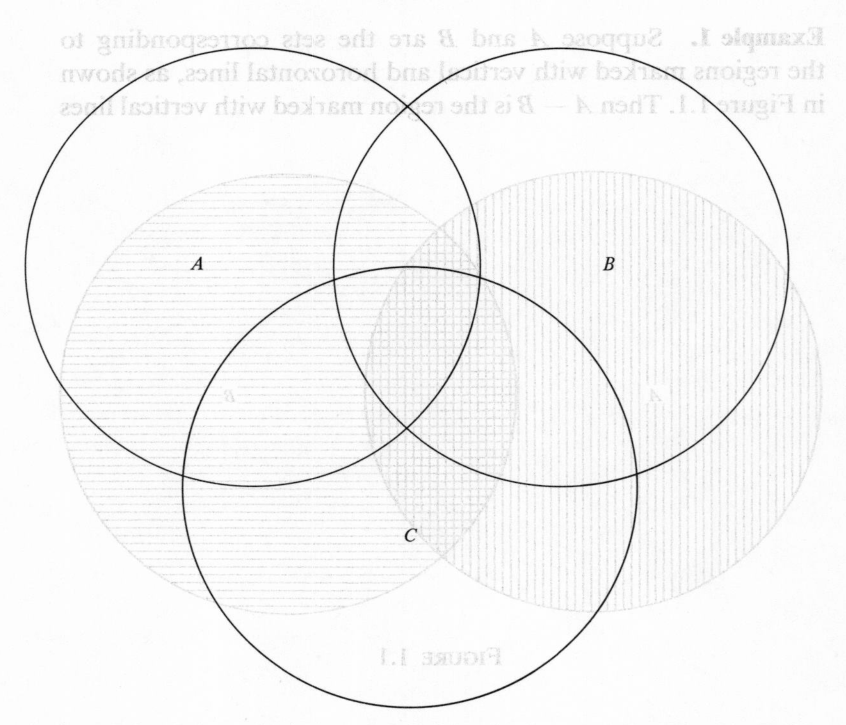

FIGURE 1.3

Example 2. Suppose in Figure 1.3 that A, B, and C are colored red, blue, and yellow, respectively. Then $A \cap B$ would be colored purple, $B \cap C$ green, and $C - A$ would consist of all areas colored yellow or green.

It should be obvious that even in the case of two sets, A and B, a single Venn diagram cannot illustrate all the possible ways in which the sets are related. We may have $A = B$, $A \subset B$, $B \subset A$, $A \cap B = \varnothing$, or none of these, and the diagram is different in each case. We urge that Venn diagrams be used, but with the understanding that they are imperfect sketches and generally do not illustrate some configurations which can occur in a given situation.

Beginning with the following exercises the reader will be doing set theory in earnest. Several of the exercises deal with the ways union, intersection, and complementation interact. Such problems are essentially computational and practically self-proving once the basic techniques have been acquired. We work out a simple problem to show what a proof in set theory looks like.

Problem. Given sets A, B, and C prove that $(A - B) - C = A - (B \cup C)$.

We shall write out a shorthand version of the proof, that is, one making liberal use of symbolism. Most people think in longhand but, in doing simple set theory problems, write in shorthand. Too much symbolism is a bad thing, so we urge the reader to use it sparingly once he has mastered it.

PROOF

	$x \in (A - B) - C$	
$\Leftrightarrow$	$x \in A - B$ and $x \notin C$	(definition of complement)
$\Leftrightarrow$	$x \in A$ and $x \notin B$ and $x \notin C$	(definition of complement)
$\Leftrightarrow$	$x \in A$ and $x \notin (B \cup C)$	(definition of union)
$\Leftrightarrow$	$x \in A - (B \cup C)$	(definition of complement).

Figure 1.4 is a Venn diagram illustrating the statement we have proved.

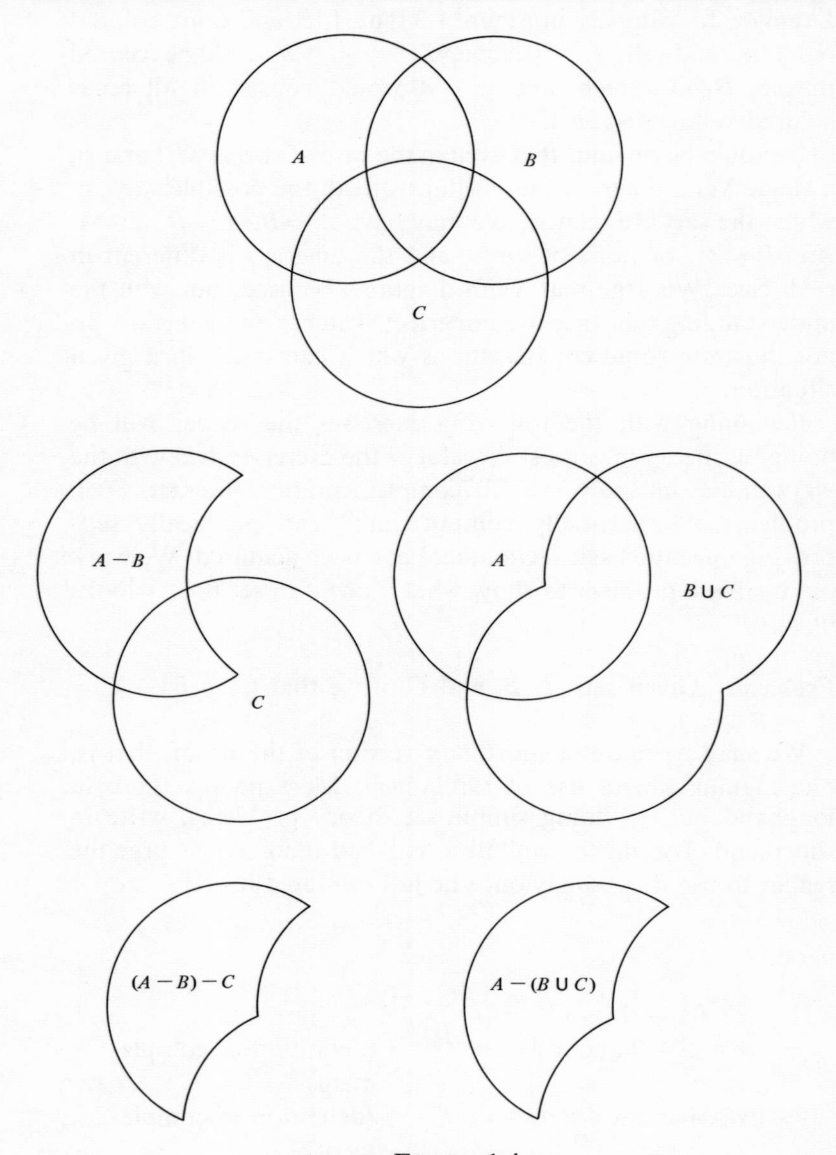

FIGURE 1.4

EXERCISES

The following list of sets will be used in Exercises 1 and 2. Here all objects under consideration are real numbers.

$$A = \{x \mid 5 \leq |x|\},$$
$$B = \{x \mid x^2 + 6 = 7x\},$$
$$C = \varnothing,$$
$$D = \{x \mid \cos x\pi = 1\},$$
$$E = \{x \mid -7 \leq x \leq 3\}.$$

1. Compute the following intersections, unions, and differences. For example, $A \cap B = \{6\}$.

$$\begin{array}{lll} D \cap B & A \cup B & C - A \\ D \cap E & B \cup E & B - D \\ B \cap E & D \cup B & D - A \\ A \cap E & A \cup E & A - E \end{array}$$

2. Solve the following "equations." Here X, the solution, is to be one of the above sets.

$$X - \{1\} = \{6\},$$
$$X - \{x \mid |x| < 5\} = \{-5, 5\},$$
$$X \cap \{x \mid -6 \leq x \leq 4\} = \{x \mid -6 \leq x \leq 3\},$$
$$X \cap \{x \mid -6 \leq x \leq 4\} = \{-6, -4, -2, 0, 2, 4\}.$$

3. The operations of union and intersection behave somewhat like addition and multiplication. For example, it is clear that each operation satisfies the "commutative" and "associative" laws: $A \cup B = B \cup A$, $(A \cup B) \cup C = A \cup (B \cup C)$, and similarly for intersection. Show that the following "distributive laws" also hold.

(a) $A \cup (B \cap C) = (A \cup B) \cap (A \cup C)$.
(b) $A \cap (B \cup C) = (A \cap B) \cup (A \cap C)$.

4. Prove the following two rules, called the *DeMorgan formulas*:

(a) $A - (B \cup C) = (A - B) \cap (A - C)$
(b) $A - (B \cap C) = (A - B) \cup (A - C)$

5. The purpose of this problem is to emphasize that the Cartesian product is fundamentally different from the other three operations we have introduced. Begin by verifying the following propositions. Here A and B are any two sets.

(a) $A \subseteq A \cup B$, and $A = A \cup B \Leftrightarrow B \subseteq A$.
(b) $A \cap B \subseteq A$, and $A \cap B = A \Leftrightarrow A \subseteq B$.
(c) $A - B \subseteq A$, and $A - B = A \Leftrightarrow A \cap B = \varnothing$.

Thus some sort of containment relation holds between the new set and at least one of the old ones.

(d) Complete and prove the following statement: $A \cap B = A \cup B \Leftrightarrow$ ———.

(e) Give an example of two sets A and B such that $A \times B$ does not contain nor is contained in either of A, B.

(f) Give an example of a nonempty set A such that

$$A \times A \subseteq A.$$

(g) Suppose A has m elements and B has n elements. Give the best answer you can to the following questions (there is an exact answer in only one case).
How many elements does $A \cap B$ have?
How many elements does $A \cup B$ have?
How many elements does $A - B$ have?
How many elements does $A \times B$ have?

7. COLLECTIONS OF SETS

Frequently it is necessary to consider sets whose elements are themselves sets; usually in such a case we shall speak of a *collection* or *family* of sets. The object of this section is to extend the operations of union and intersection to collections of sets.

Let $\mathscr{A}$ denote a collection of sets. The *union* of the collection $\mathscr{A}$, denoted $\cup\mathscr{A}$, is the set of all objects which belong to at least one member of $\mathscr{A}$, and the *intersection* of the collection $\mathscr{A}$, denoted $\cap\mathscr{A}$, is the set of all objects which belong to every member of $\mathscr{A}$.

Notice that if $\mathscr{A}$ has just two elements, A and B, then $\cup\mathscr{A} = A \cup B$ and $\cap\mathscr{A} = A \cap B$. Thus the new concepts of union and intersection are generalizations of the old.

The brace notation can be very useful in defining collections of sets. For example, suppose $\mathscr{A}$ and $\mathscr{B}$ are collections of sets; then the collection $\mathscr{C}$ obtained by taking the complement of each element of $\mathscr{B}$ in each element of $\mathscr{A}$ can briefly and clearly be written

$$\mathscr{C} = \{A - B \mid A \in \mathscr{A}, B \in \mathscr{B}\}.$$

We now give some examples. The reader should verify these carefully before going on.

Example. Let X be a set and let $\mathscr{X}$ be the collection of all subsets of X. Then $\cup\mathscr{X} = X$ and $\cap\mathscr{X} = \varnothing$.

Example. Let $\mathscr{C}$ denote the collection of all nondegenerate circles in the plane which are centered at the origin O. Let $\mathscr{D}$ be the collection of all such circles whose radius is a rational number. Then $\cap\mathscr{C} = \cap\mathscr{D} = \varnothing$, $\cup\mathscr{C}$ is the set of all points of the plane except O, and $\cup\mathscr{D}$ is the set of all points in the plane whose distance from O is a rational number.

We now list some useful facts about unions and intersections. It is instructive to try to find analogies for these among the exercises of Section 6. Here script letters denote collections of sets.

7.1. $\cap\mathscr{A} \subseteq \cup\mathscr{A}$.

7.2. For each A in $\mathscr{A}$, $A \subseteq \cup\mathscr{A}$ and $\cap\mathscr{A} \subseteq A$.

7.3. Distributive laws:

(a) $(\cup\mathscr{A}) \cap (\cup\mathscr{B}) = \cup\{A \cap B \mid A \in \mathscr{A}, B \in \mathscr{B}\}$.
(b) $(\cap\mathscr{A}) \cup (\cap\mathscr{B}) = \cap\{A \cup B \mid A \in \mathscr{A}, B \in \mathscr{B}\}$.

7.4. DeMorgan formulas:

(a) $X - \cup\mathscr{A} = \cap\{X - A \mid A \in \mathscr{A}\}$.
(b) $X - \cap\mathscr{A} = \cup\{X - A \mid A \in \mathscr{A}\}$.

As an illustration we prove one of the distributive laws, 7.3(b). We assume that $\mathscr{A} \neq \varnothing$ and $\mathscr{B} \neq \varnothing$, since the result is trivial otherwise.

Let L denote the set $(\cap\mathscr{A}) \cup (\cap\mathscr{B})$ and let R denote the set $\cap\{A \cup B \mid A \in \mathscr{A}, B \in \mathscr{B}\}$.

To prove $L = R$ we first prove that $L \subseteq R$. Suppose, then, that x is in L. Thus $x \in \cap\mathscr{A}$ or $x \in \cap\mathscr{B}$. If $x \in \cap\mathscr{A}$, then for every A in $\mathscr{A}$ and every B in $\mathscr{B}$ we have $x \in A \subseteq A \cup B$; hence x belongs to $\cap\{A \cup B \mid A \in \mathscr{A}, B \in \mathscr{B}\} = R$. A similar argument applies if $x \in \cap\mathscr{B}$. Thus, in either case, if x is in L, then x is in R and therefore $L \subseteq R$.

We next show that if x is not in L, then x is not in R; by the remark at the end of Section 4, this implies that $L \supseteq R$, which together with the containment proved above shows that $L = R$. So, suppose that x is not in L; then x does not belong to either $\cap\mathscr{A}$ or $\cap\mathscr{B}$. Hence there exist elements A in $\mathscr{A}$ and B in $\mathscr{B}$ such that $x \notin A$ and $x \notin B$; but then $x \notin A \cup B$ and therefore, by definition of intersection, x is not in R.

EXERCISES

1. Prove facts 7.1 and 7.2.
2. Prove fact 7.3(a).
3. Prove the DeMorgan formulas.
4. For each real number r, let A_r denote the set of real numbers x such that $-r \leq x \leq r$. Let $\mathscr{A}$ be the collection of all such sets; that is, $\mathscr{A} = \{A_r \mid r \text{ is a real number}\}$. Describe the sets $\cap\mathscr{A}$ and $\cup\mathscr{A}$.
5. Find a collection $\mathscr{A}$ of nonempty subsets of the set R of real numbers such that $\cap\mathscr{A} = \varnothing$ and yet given any two elements A and B of $\mathscr{A}$ either $A \subseteq B$ or $B \subseteq A$. Can such a collection have only finitely many terms?

CHAPTER 2

Functions

122296

1. INTRODUCTION

It is possible that the reader was introduced to the concept of a function in the following way.

A function from a set A into a set B is a rule which assigns to each element of A one and only one element of B. Suppose the letter f denotes a function from A into B; if a is a member of A, then the unique element of B which is assigned to a by f is called the value of f at a or the image of a under f and is denoted simply $f(a)$.

The graph of f is the set of all elements (a, b) of $A \times B$ such that $b = f(a)$. Notice that if the graph of a function is known, then the function is known; to put this more precisely, if it is known exactly what elements of $A \times B$ are members of the graph of f, then it is known what value f assigns to each element of a.

Except for the lamentable fact that we do not know exactly what we are talking about (since the words "rule" and "assign" have not been defined), this is a very good formulation of the concept of a function. The intuitive content is so well understood and so useful that much of elementary mathematics can be done without being any more precise. Unfortunately at a more advanced level, serious problems arise if the notion of a function is left undefined; to do some kinds of mathematics it is absolutely necessary to know what is a function and what is not.

So we are faced with the problem of finding a set theoretic definition of the word "function." Clearly not just any definition will do; for example, everyone would object to a definition formulated in such a way that half the calculus becomes invalid. Thus the problem is to find a precise way of formulating the concept of a function which conforms as nearly as possible with our intuitive notion of what a function is.

Happily this problem has a very neat solution, which is given below. This may be better appreciated by those who first make a reasonable effort to formulate their own definitions.

2. BASIC DEFINITIONS

A *function* from a set A into a set B is a subset F of $A \times B$ such that if a is an element of A, then there is one and only one element b of B such that (a, b) is an element of F.

Remark. We have used a capital letter to denote a function in the above definition merely to emphasize that a function is a set. From now on we will follow the general custom and use lower-case letters, usually $f, g, h, \ldots$, to denote functions.

The connection between the intuitive notion of a function and the definition is simple. What we have defined as a function is what was formerly called the graph of a function. Our earlier remarks concerning graphs should indicate that the intuitive content has not been changed; it has merely been made precise.

To save words we shall indicate that f is a function from A into B by writing simply $f: A \to B$.

We now quickly introduce some terminology which is useful in discussing functions. In what follows, f denotes a function from a set A into a set B.

If a is in A, then the unique element b of B such that (a, b) is an element of f will be called the *image* of a and denoted by $f(a)$.

The *domain* of f is the set A and the *range* of f is the set of all points of B which are images under f of points of A; that is, the range of f is the set $\{b \in B \mid \text{for some } a \in A, b = f(a)\}$.

The function $f: A \to B$ is *one to one*, written 1–1, provided that if a and a' are in A and $a \neq a'$, then $f(a) \neq f(a')$.

If $f: A \to B$ and C is a subset of B, then we say f is *onto* C in case every element of C is the image of a point in A under f. Thus f is onto C if and only if C is a subset of the range of f.

Example 1. Let $A = B =$ the set R of real numbers and let $f: A \to B$ be the function whose value at $a \in A$ is the element a^2 of B. [In the cross-product notation we might write

$$f = \{(a, b) \in A \times B \mid b = a^2\}.]$$

Then the range of f is $\{b \in B \mid b \geq 0\}$, and f is not 1–1.

Example 2. Let $\mathscr{P}(X)$ denote the collection of all subsets of a set X and let $f: \mathscr{P}(X) \to \mathscr{P}(X)$ be the function whose value at $A \in \mathscr{P}(X)$ is the set $X - A$. Then f is 1–1 from $\mathscr{P}(X)$ onto $\mathscr{P}(X)$.

Example 3. If X is any set, then the function $\{(x, x) \mid x \in X\}$ is called the *identity function* of X. The identity function is 1–1 from X onto X.

When it is convenient we shall follow the practice of defining a function by specifying its domain and the image of each element of the domain. Frequently the image can be specified by a formula. For example, the functions given above can be described briefly as follows:

2.1. $f\colon I \to I, f(a) = a^2$.
2.2. $f\colon \mathscr{P}(X) \to \mathscr{P}(X), f(A) = X - A$.
2.3. $f\colon X \to X, f(x) = x$.

EXERCISES

In the following problems, R denotes the set of real numbers.

1. Tell which of the following subsets of $R \times R$ are functions from R into R and sketch these sets (that is, sketch graphs).

(a) $\{(a, b) \mid a^2 = b\}$
(b) $\{(x, y) \mid x^2 = y^2\}$
(c) $\{(s, t) \mid t = 1\}$
(d) $\{(a, b) \mid A = 1\}$
(e) $\{(x, y) \mid y = \sqrt{x}\}$
(f) $\{(x, y) \mid xy = 1\}$
(g) $\{(x, y) \mid x^y = 2\}$
(h) $\{(a, b) \mid \text{if } -1 \leq a < 1, \text{ then } b = a^2, \text{ and } b = 0 \text{ otherwise}\}$
(i) $\{(x, y) \mid x^2 = y^2\} \cap \{(x, y) \mid y \geq 0\}$
(j) $\{(x, t) \mid x \geq t\}$

2. In each of the following you are given a formula which is to be used to define a function whose domain is a subset of R and whose range is contained in R. Specify the "largest possible" domain in each case. For example, corresponding to $f(x) = \sqrt{x}$ the largest possible domain is $\{x \mid 0 \leq x\}$.

(a) $g(a) = a^{1/3}$
(b) $h(y) = 1/y$
(c) $i(x) = \dfrac{x^2 + 1}{x^2 - 1}$
(d) $f(z) = \dfrac{z^2}{z^2 - 2}\sqrt{z + 1}$
(e) $g(x) = \sin^{-1} x$
(f) $h(y) = \ln(1 + y)$

3. In each of the following given an example of a function $f\colon R \to R$ with the property stated.

(a) The range of f is $(-1, 1)$ and f is 1–1.
(b) Each element of R is the image under f of exactly two elements of R.

(c) The range of f is the set $\{x \in R \mid x \text{ is not an integer}\}$ and f is 1–1.

4. For this problem you may need to review some calculus—particularly the fundamental theorem of calculus, which states the relationship between integrals and derivatives. Let $\mathscr{D}$ be the set of all functions from R into R having continuous derivatives and let $\mathscr{C}$ be the collection of all continuous functions from R into R.

(a) Is $\{(f, g) \in \mathscr{C} \times \mathscr{D} \mid f = g'\}$ a 1–1 function from $\mathscr{C}$ into $\mathscr{D}$? Justify your answer.
(b) Same question for $\{(f, g) \in \mathscr{D} \times \mathscr{C} \mid f = g'\}$.
(c) Let $\mathscr{P}(\mathscr{D})$ be the collection of all subsets of $\mathscr{D}$. Consider the set $\{(f, A) \in \mathscr{C} \times \mathscr{P}(\mathscr{D}) \mid g \in A \Leftrightarrow g' = f\}$. Show that this is a function from $\mathscr{C}$ into $\mathscr{P}(\mathscr{D})$. Is it 1–1? Onto? What is another name for the image under this function of an element f of $\mathscr{C}$?

3. MORE DEFINITIONS

If $f\colon A \to B$ is 1–1 and onto, then we can define, in a natural way, a function from B to A which, in a crude sense, reverses f. To be precise, suppose f is as above and consider the set $\{(b, a) \in B \times A \mid b = f(a)\}$. Since f is 1–1 and onto there is, for each b in B, one and only one element a in A such that $b = f(a)$; hence the above set is a function from B into A. This function is 1–1 because f is a function (no two points in B are the images of the same point in A). Since f has A as its domain, the new function is onto A. The function induced by f in the above way is called the *inverse* of f and is denoted f^{-1}.

The reader is cautioned that f^{-1} should be thought of as one symbol; the minus one is not an exponent and it is dangerous nonsense to read f^{-1} as "f to the minus one."

Here are some examples of functions and their inverses.

3.1. The identity function on any set is its own inverse.

3.2. The function given in Example 2 of Section 7, $f\colon \mathscr{P}(X) \to \mathscr{P}(X)$, $f(A) = X - A$, is its own inverse.

3.3. Let f be the exponential function, $f(x) = e^x$, defined on R, the set of real numbers. Then f is 1–1 and onto R^+ and f^{-1} is given by $f^{-1}(x) = \ln x$.

There is a natural way of combining old functions to get new ones. Suppose $f\colon A \to B$, $g\colon C \to D$, where B is a subset of C.

The function $h: A \to D$ defined by $h(a) = g(f(a))$ is called the *composition* of f and g, or just the composite function, and denoted $g \circ f$. [The letter f is on the right because f attacks first: $g \circ f(a) = g(f(a))$.]

Observe that the formula used to define $g \circ f$ makes no sense unless the range of f is a subset of the domain of g; if this condition is not satisfied the composite function does not exist.

In the following examples R denoted the set of real numbers and R^+ the set of nonnegative real numbers. We consider three functions:

$$\begin{aligned} &f: R \to R, && f(x) = x^2, \\ &g: R^+ \to R, && g(x) = \sqrt{x} \text{ (positive square root)}, \\ &h: R \to R, && h(x) = -x. \end{aligned}$$

3.4. $g \circ f: R \to R$ exists and, in fact, $g \circ f$ is the absolute value function.

3.5. $f \circ g: R^+ \to R$ exists and is the identity function on R^+.

3.6. $g \circ h$ does not exist.

3.7. $h \circ g: R^+ \to R$ exists; the image of $x \in R^+$ under this function is the negative square root of x.

3.8. $h \circ h$ (exists and) is the identity function on R.

The following observations about inverses follow easily from the definitions.

Remark on Inverses. Suppose $f: A \to B$ *is* 1–1 *and onto. Then*
1. $(f^{-1})^{-1} = f$.
2. $f^{-1} \circ f$ *is the identity on* A.
3. $f \circ f^{-1}$ *is the identity on* B.

We end this section by introducing a useful bit of notation. If $f: A \to B$ is a function and C is a subset of A, then by $f(C)$ we mean the set of images of the elements of C; that is, $f(C) = \{f(x) \mid x \in C\}$. Exercise 4 is designed to familiarize the reader with this notation.

EXERCISES

1. Show that the composition of 1–1 maps is 1–1.

2. Suppose $f: A \to B$ is onto and that $g: B \to A$ is a function such that $g \circ f: A \to A$ is 1–1. Prove that f and g are both 1–1.

3. Let A be the circle $x^2 + y^2 = 1$ in the plane. Find a function $f: A \to A$ which is not the identity on A such that $f \circ f \circ f$ is the identity on A.

4. Let $f: A \to B$ be a function. Show that if f is 1–1 and C is a subset of A, then $f^{-1} \circ f(C) = C$. State the corresponding result when f is not assumed 1–1 and either prove it or give a counterexample.

Show that if C and D are subsets of A and f is 1-1, then $f(C \cup D) = f(C) \cup f(D)$ and $f(C \cap D) = f(C) \cap f(D)$. Show that the latter is not true (in general) if f is not 1–1.

5. The following result will be used in Chapter 3. Let n be some positive integer and let A be the set $\{1, 2, \ldots, n\}$. Let m be an element of A and define a function $g: A - \{m\} \to A$ by

$$g(x) = \begin{cases} x & \text{if } x < m \\ x - 1 & \text{if } x > m. \end{cases}$$

Show that g is 1–1 from $A - \{m\}$ onto the set $\{1, 2, \ldots, n - 1\}$.

6. Let A denote the set of real numbers x such that $0 \leq x \leq 1$. Figure 2.1 shows two functions f and g from A onto A.

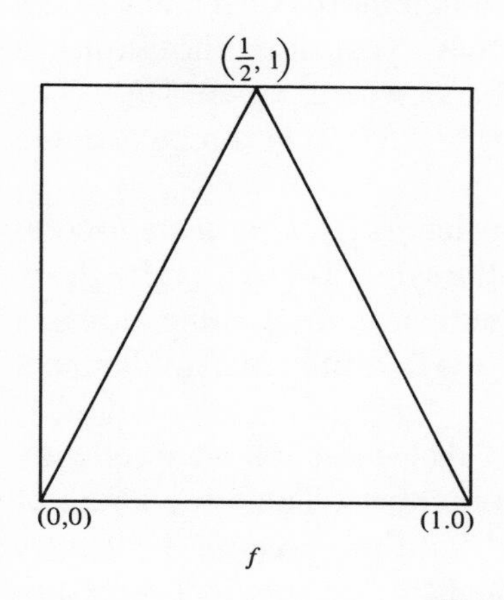

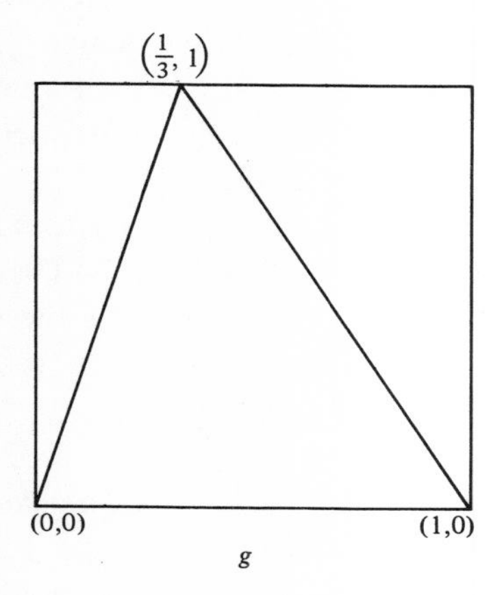

FIGURE 2.1

(a) Write out formulas which define f and g. (*Hint*: Each function requires two formulas.)

(b) Find a function $h: A \to A$ which is 1–1 and onto such that $f = g \circ h$.

4. SEQUENCES

In this and the next section we introduce the concepts of sequence and subsequence which are used extensively in studying the real number system.

Let A be a set; a *sequence* in A is a function $x: I^+ \to A$. (Here, and from now on, I^+ denotes the set of positive integers.) The ith *term* of the sequence is the image of the integer $i \in I^+$; thus the first term of $x: I^+ \to A$ is $x(1)$, the third term is $x(3)$, and the nth is $x(n)$. Usually when dealing with sequences we use *subscripts* rather than the functional notation to indicate which term is under consideration; that is, we write x_i rather than $x(i)$ to denote the ith term.

Since a sequence, by definition, has I^+ as its domain, we can specify a sequence merely by specifying what each term is. Quite often this can be done by a single formula; if this is the case, then the formula, enclosed in braces, is used to denote the sequence.

For example, $\{1/n\}$ denotes the sequence whose nth term is $1/n$, for every $n \in I^+$. Clearly it is much more convenient to say "consider the sequence $\{1/n\}$" than "consider the sequence $x: I^+ \to R$ (the set of real numbers) where $x(n) = 1/n$." The sequence $x: I^+ \to R$ defined by $x(i) = 2^i/(3i+1)$ can be denoted simply $\{2^i/(3i+1)\}$.

We remark that, customarily, the letters i, j, k, or n are used to denote elements of I^+. Therefore, the symbols $\{1/i\}$, $\{1/j\}$, $\{1/k\}$, and $\{1/n\}$ all denote the same sequence. In writing down a sequence, each of us is free to use his favorite letter to denote a typical element of I^+.

If $\{x_n\}$ is a sequence in the set of real numbers, we frequently say that $\{x_n\}$ is a sequence *of* real numbers. This is an abuse of terminology but it does reflect the fact that, because of the way they are used, sequences are thought of as sets indexed (subscripted) by integers. Similarly, if $\{x_n\}$ is a sequence whose terms are integers, functions, sets, and so on, we say that $\{x_n\}$ is a sequence *of* integers, functions, sets, and so on.

Here are some examples of such sequences. Let F denote the set of all functions from the set $\{x \in R \mid -1 \leq x \leq 1\}$ into itself. Let $\{f_i\}$ be the sequence in F whose ith term is the function defined by $f_i(x) = x^i$. Thus the first term of $\{f_i\}$ is the identity function, the second term is the function $f(x) = x^2$, and so on. Figure 2.2 shows the first few terms of this sequence of functions.

Let $\{L_i\}$ be the sequence whose ith term is the set of points (x, y) in the plane satisfying $x = 1/i$, $0 \leq y \leq 1/i$. Then $\{L_i\}$ is a sequence of vertical-line segments (Figure 2.3).

EXERCISES

1. Write the following sequences in the brace notation. Some thought may be required to get a single formula in each case.

(a) $x: I^+ \to I$; $x(n) = \begin{cases} 1 & \text{if } n \text{ is even,} \\ -1 & \text{if } n \text{ is odd.} \end{cases}$

(b) $x: I^+ \to I^+$; $x(n) = \begin{cases} 1 & \text{if } n \text{ is even,} \\ 0 & \text{if } n \text{ is odd.} \end{cases}$

(c) $x: I^+ \to I$, where $x(1) = 1$ and, for $n \geq 2$, $x(n) = x(n-1) + 1$.

(d) $x: I^+ \to I$, where $x(n) = \begin{cases} 1 & \text{if } n = 1, 5, 9, \ldots, \\ 0 & \text{if } n = 2, 6, 10, \ldots, \\ -1 & \text{if } n = 3, 7, 11, \ldots, \\ 0 & \text{if } n = 4, 8, 12, \ldots. \end{cases}$

2. Exhibit sequences whose first few terms are those given below There is no unique answer in any case.

(a) $\frac{1}{2}, \frac{1}{4}, \frac{1}{8}, \frac{1}{16}, \ldots$

(b) $2, \frac{5}{2}, \frac{10}{3}, \frac{17}{4}, \ldots$

(c) $2, 8, 32, 128, \ldots$

3. Define a sequence whose range is the set of all integers.

4. Draw a picture of the first few terms of the sequence $\{C_i\}$, where C_i is the set of points (x, y) in the plane satisfying $(x - i)^2 + y^2 = (1/i)^2$.

5. Find a sequence $\{x_i\}$ of real numbers, no two terms of which are equal, with the property that if x_n and x_m are any two terms of the sequence with $x_n < x_m$ then there is a third term x_k such that $x_n < x_k < x_m$.

6. Let $\{A_i\}$ be a sequence of subsets of some set X. Then for each integer m the symbol $\bigcap_{i \geq m} A_i$ has the obvious meaning; it is the set of all points of X which belong to A_i for all $i \geq m$. Similarly $\bigcup_{i \geq m} A_i$ is the set of points belonging to some A_i where $i \geq m$.

If $m > n$, what containment relation holds between $\bigcap_{i \geq m} A_i$ and $\bigcap_{i \geq n} A_i$? Between $\bigcup_{i \geq m} A_i$ and $\bigcup_{i \geq n} A_i$?

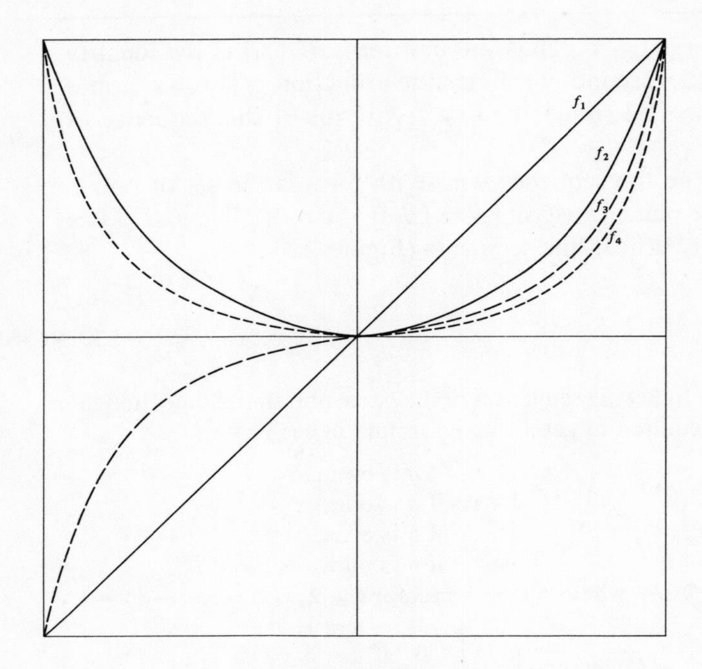

FIGURE 2.2

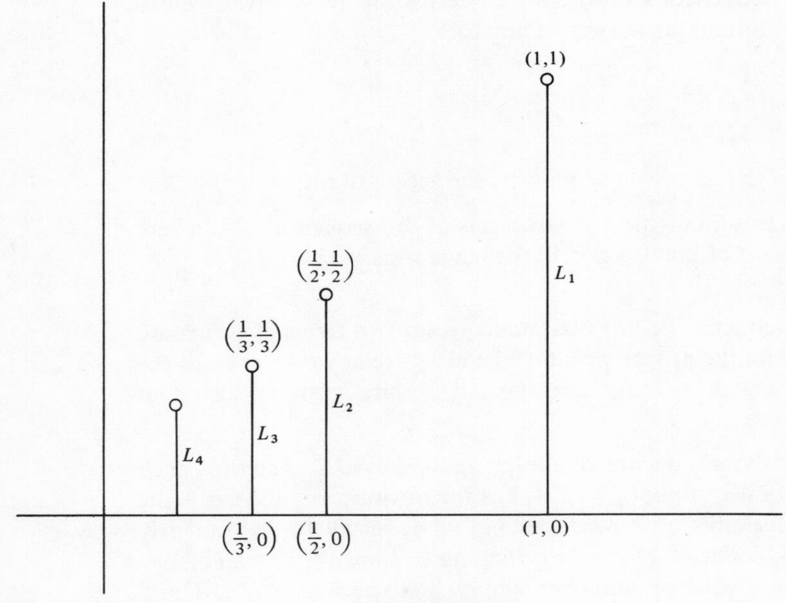

FIGURE 2.3

7. Let $\{A_i\}$ be a sequence of subsets of a set X and define a new sequence $\{\tilde{A}_i\}$ by setting $\tilde{A}_i = A_1 \cap A_2 \cap \ldots \cap A_i$ for each i. Verify the following results.
(a) $\tilde{A}_i \subseteq A_i$, for all i.
(b) $\{\tilde{A}_i\}$ is a *decreasing* sequence, that is, $\tilde{A}_i \supseteq \tilde{A}_{i+1}$, for all i.
(c) $\bigcap_{i\geq 1} \tilde{A}_i = \bigcap_{i \geq 1} A_i$.
Is it true that $\bigcup_{i\geq 1} \tilde{A}_i = \bigcup_{i\geq 1} A_i$?

8. Let $\{A_i\}$ be a sequence of subsets of a set X. For each positive integer j let $B_j = \bigcup_{i\geq j} A_i$ and $C_j = \bigcap_{i \geq j} A_i$. Define two sets A^u and A^l by $A^u = \bigcap_{j\geq 1} B_j$ and $A^l = \bigcup_{j \geq 1} C_j$. Show that a point x of X belongs to A^u if and only if x belongs to A_i for infinitely many values if i. Formulate and prove the corresponding statement for A^l.

9. Continuing the previous problem, let $\{A_i\}$ be the sequence of subsets of R (the set of real numbers) defined by letting A_i be the set of numbers whose distance from one of the integers $1, 2, \ldots, i$ is less than $1/i$. Compute $\bigcap_{i\geq 1} A_i$, $\bigcup_{i\geq 1} A_i$, A^u, and A^l.

5. SUBSEQUENCES

Frequently, given a sequence $\{x_n\}$, we wish to form a new sequence by deleting some of the terms of $\{x_n\}$ and keeping the rest. For example, let $\{f_i\}$ be the sequence of functions given in Section 4. Suppose, for some reason, we wish to form a sequence by deleting those functions and only those which take on negative values. One way to do this would be to select f_2 as the first term of the new sequence, f_4 as the second term, f_6 as the next, and so on. There are, of course, other ways to form the new sequence; we could select f_4 as the first term of the new sequence, f_2 as the second, f_8 as the next, f_6 as the next, and continue alternating this way. Here we have not only deleted terms of $\{f_i\}$ but we have changed the order of appearance of those terms which remain. For the purposes we have in mind it is inconvenient to allow what we have just described to happen, so it will be made illegal by our definition.

Let $\{x_n\}$ and $\{y_n\}$ be sequences in a set A, say $x_n = x(n)$, where $x\colon I^+ \to A$, and $y_n = y(n)$, where $y\colon I^+ \to A$. We say that $\{y_n\}$ is a subsequence of $\{x_n\}$ provided that there is a function $f\colon I^+ \to I^+$ such that

1. $y = x \circ f$.
2. f is *order preserving* in the sense that if $n < m$, then $f(n) < f(m)$.

Let us examine the definition. The first requirement, $y = x \circ f$, can be rewritten as follows: For each n, $y(n) = (x \circ f)(n) = x(f(n))$; and in our subscript notation this says $y_n = x_{f(n)}$. In words, the nth term of the subsequence is the $f(n)$th term of the original sequence. Let us call f the *selection function* because it tells us what terms $\{$of $\{x_n\}$ to select to get $y_n\}$.

The second requirement on f ensures that those terms of the original sequence which are selected for the subsequence appear in the right order. If y_i and y_j are terms of the subsequence with $i < j$, then $y_i = x_{f(i)}$ and $y_j = x_{f(j)}$ and $f(i) < f(j)$, so that these terms appear in the same order in both sequences.

Here are some examples.

5.1. Sequence: $\{1/n\}$
Selection function: $f(n) = n + 3$
Subsequence: $\{1/(n+3)\}$

5.2. Sequence: $\{n^2\}$
Selection function: $f(n) = n^2 + 2$
Subsequence: $\{n^4 + 4n^2 + 4\}$

5.3. Given any sequence $\{x_n\}$, suppose we wish to form the subsequence consisting of those terms of $\{x_n\}$ having even subscripts. The selection function which does this is the function $f(n) = 2n$ and the corresponding subsequence is $\{x_{2n}\}$. The first term of this subsequence is x_2, the second is x_4, and so on.

5.4. Consider any sequence $\{x_n\}$ and let $\{y_n\}$ be the sequence whose first term is x_2, whose second is x_1, and whose nth term, for $n \geq 3$, is x_n. Then $\{y_n\}$ is not a subsequence of $\{x_n\}$ because the only possible selection function is $f(1) = 2, f(2) = 1, f(n) = n (n \geq 3)$; and this is not order preserving.

We note the following fact about subsequences: *If $\{z_n\}$ is a subsequence of $\{y_n\}$ and $\{y_n\}$ is a subsequence of $\{x_n\}$, then $\{z_n\}$ is a subsequence of $\{x_n\}$*, or, to put it more briefly, a subsequence of a subsequence is a subsequence.

To prove this, let f and g be order-preserving functions from I^+ into I^+ such that $z_n = y_{f(n)}$, for every $n \in I^+$, and $y_n = x_{g(n)}$, for every $n \in I^+$. Let $h = g \circ f$; then h is order preserving

(Exercise 1) and, for every $n \in I^+$, we have

$$z_n = y_{f(n)} = x_{g(f(n))} = x_{h(n)}.$$

EXERCISES

1. Prove that the composition of order-preserving functions from I^+ into I^+ is order preserving.

2. Show, say by induction, that if $f\colon I^+ \to I^+$ is order preserving, then, for every $n \in I^+$, $n \leq f(n)$. Is the converse true?

3. Given a sequence $\{x_n\}$, what selection function yields the subsequence whose terms are the terms of $\{x_n\}$ with odd subscripts?

4. Given $\{x_n\}$, what selection function yields the subsequence $\{y_n\}$ such that $y_1 = x_1$, $y_2 = x_6$, $y_3 = x_{11}$, $y_4 = x_{16}$, and so on?

5. Let $\{x_n\}$ be a sequence of real numbers and suppose the range X of $\{x_n\}$ is finite. Show that $\{x_n\}$ has a constant subsequence. Show that if X is infinite, then $\{x_n\}$ has a subsequence no two terms of which are equal.

6. In each part of this problem you are given a sequence and a selection function. Write out the numerical values of the first four terms of the corresponding subsequence.

(a) $x_n = 1/2n$, $f(n) = n + 2$
(b) $x_n = 1/2n$, $f(n) = 2n - 1$
(c) $x_n = (-1)^n$, $f(n) = 2n + 5$
(d) $x_n = (-1)^n + 1$, $f(n) = n(n + 1)/2$
(e) $x_n = \cos n\pi/n$, $f(n) = 2n$
(f) $x_n = \cos n\pi/n$, $f(n) = 2n + 1$

7. Let $\{A_n\}$ be the sequence of subsets of R (the set of real numbers) defined in Exercise 9 of the preceding section. Let $\{B_n\}$ be the subsequence corresponding to the selection function $f(n) = 2n$; compute B^u and B^l. Do the same problem using the selection function $f(n) = n + 5$.

CHAPTER 3

Relations

1. INTRODUCTION

To analyze a situation involving more than a few objects (variables, data, alternatives, or whatever) it is extremely convenient to establish a basis for comparing these objects and to classify them, or at least differentiate between them, on this basis. Indeed, this may be more than convenience; it might be argued that in a real-life situation a person cannot concentrate his attention on a set of objects without beginning immediately to compare them in one or more ways. Of course, the basis and extent of comparison will be tailored to suit the objects under study and the objectives of the study itself.

For example, given a set of coins we may wish to compare them on the basis of weight, face value, value of the metal they contain, value as collectors items, or perhaps on some aesthetic criterion. Somewhat more abstractly, given a collection of problems to be resolved one might deal with them in the chronological order of their presentation, in the decreasing order of importance, or perhaps in increasing order of difficulty. Each of the above involves setting up what is called an order relation among the objects. A precise definition will be given presently.

Another type of comparison involves the notion of equivalence. Here one decides that, for the purpose at hand, objects will be considered the same (that is, equivalent) if they share certain properties, and different if they do not. For example, we have all experienced the moment of truth in front of a coin-operated device when it becomes clear that (at that instant) ten dimes, twenty nickels, or four quarters are all equivalent but none of these is equivalent to a dollar bill.

In this chapter we shall formalize the notion of "relation," concentrating our attention on order and equivalence relations. The treatment consists mainly of definitions and examples, although there are a few theorems. The ideas introduced here permeate the rest of this book and, indeed, the rest of mathematics in general. By employing these notions efficiently one can frequently simplify a complicated situation by identifying and eliminating superficial differences.

2. RELATIONS

Given two sets X and Y we define a *relation from X to Y* to be a subset R of $X \times Y$. The *domain* of R is the set of x in X such that for some y in Y, (x, y) is an element of R. If $X = Y$ we say R is a relation *in* X, and if, moreover, the domain of R is all of X, we say R is a relation *on* X.

If X and Y are any sets, then the set $R = X \times Y$ is a relation called the *universal relation*, and the empty set, considered as a subset of $X \times Y$, is a relation called the *empty relation*. Any function from X into Y is a relation from X to Y with domain X. Given a set X, the *diagonal* Δ of $X \times X$ is the set $\{(x, x) \mid x \in X\}$. The diagonal is a relation on X; in fact, it is just the relation of equality: (x, y) is in Δ if and only if $x = y$. Since any subset of $X \times Y$ is a relation from X to Y, examples of relations abound. We shall defer giving any more examples until we can make them complicated enough to be interesting.

For most purposes the notation given above for relations is cumbersome and the following conventions are used. First, rather than write $(x, y) \in R$, where R is a relation, we write xRy. This is read "x is R-related to y" (or just "x is related to y," if there is only one relation under consideration). Most relations which occur in a natural way are denoted by symbols rather than by letters and the symbols are themselves shorthand for phrases. Common examples include the symbols $>$ and $=$. For the first of these we write "$x > y$" (rather than $(x, y) \in >$) and read "x is greater than y"; for the second, "$x = y$" and "x is equal to y."

3. PROPERTIES OF RELATIONS

Let R be a relation in a set X. We say that R is *reflexive* provided xRx for every x in the domain of R. In case R is a relation on X, R is reflexive if and only if R contains the diagonal Δ of $X \times X$.

We say that R is *symmetric* if, whenever xRy, also yRx. If the reader draws a picture of $I \times I$ where $I = \{x \mid 0 \leq x \leq 1\}$ is the unit interval of real numbers with the diagonal included, he should be able to see that a relation R on I satisfies the above condition if and only if $R \subset I \times I$ is (geometrically) symmetric with respect to Δ; hence the terminology.

Finally, R is called *transitive* if, whenever xRy and yRz, then also xRz.

Here are some important examples of relations which possess one or more of the above properties.

Example 1. Given a set X, let $\mathscr{P}(X)$ denote the collection of all subsets of X. Containment, denoted $\subseteq$, is a relation on $\mathscr{P}(X)$. This relation is reflexive and transitive but not symmetric (it is not generally true that if $A \subseteq B$, then $B \subseteq A$). The "stronger" relation of proper containment, denoted $\subset$, is transitive but not reflexive or symmetric. (Actually, the domain of $\subset$ is $\mathscr{P}(X) - \{X\}$.)

Example 2. Let R be the set of real numbers and let $>$ have the usual meaning in R; thus $x > y$ means $x - y$ is the square of a nonzero number. Then $>$ is transitive but not reflexive or symmetric. We can "make" $>$ into a reflexive relation, $\geq$, by defining $x \geq y$ to mean that $x > y$ or $x = y$. The modified relation will still fail to be symmetric.

Example 3. Let X be a plane equipped with a coordinate system and let $\mathscr{T}$ be the set of all triangles in X. Define a relation $\sim$ on $\mathscr{T}$ by $T_1 \sim T_2$ provided T_1 and T_2 are similar in the sense of plane geometry (they have the same interior angles). Then $\sim$ is reflexive, symmetric, and transitive. Congruence is another such relation on $\mathscr{T}$. On the other hand, suppose we define a relation $*$ on $\mathscr{T}$ by $T_1 * T_2$ if and only if one side of T_1 has the same length as one side of T_2. Then $*$ is reflexive and symmetric but not transitive.

Example 4. In a somewhat less mathematical vein, let H be the set of all human beings alive at a given moment. (We leave to someone else the problem of resolving the thorny biological and philosophical issues involved in making this notion precise.) Let b be the relation defined on H by xby provided x is male and x and y have the same (biological) parents; in short, xby if and only if x is the brother of y (every male being his own brother in this case). The relation b is transitive and reflexive but not symmetric. One can consider various other relations defined in H such as xay provided x is a (biological) ancestor of y; $x|y$ provided x and y have the same (biological) parents; and $x * y$ provided x and y have at least one (biological) parent in common. This is as far as we shall carry our study of human relations.

EXERCISES

In each of the following you are given one or more relations in a set. Tell what the domain of each relation is and whether it is reflexive, symmetric, and/or transitive.

1. The relations a, $\perp$, and $*$ defined in Example 4.

2. The relations $\equiv_n$ and $|$ defined in the set of integers by

(a) $x \equiv_n y$, provided $x - y$ is a multiple of n (n being a fixed integer).
(b) $x|y$, provided y is an integral multiple of x ($y = nx$, for some integer n).

3. The relations defined below on the set of all differentiable functions from R (the real numbers) into R.
(a) $f \sim g$, provided $f' = g'$.
(b) $f \circ g$, provided that for some real number t_0, depending on f and g, $f(x) \geq g(x)$ for all $x \geq t_0$. (Here $\geq$ is the usual order relationship between real numbers.)
(c) $f * g$, provided $f^{(n)} = g^{(m)}$ for some nonnegative integers n and m. (Here the exponent denotes differentiation.)
(d) $f ** g$, provided $f - g$ is a polynomial function.

4. The relations defined on the set of positive real numbers by
(a) xPy, provided $xy = 1$.
(b) xQy, provided $x^2 + y^2 = 1$.
(c) xRy, provided $x^2 + y^2 = 1$ or $x = \pm y$.

4. EQUIVALENCE RELATIONS

A relation on a set X is called an *equivalence relation* (on X) provided it is reflexive, symmetric, and transitive. The prime example of an equivalence relation is equality. In example 3 of Section 3 we noted that similarity and congruence are equivalence relations on the set of all triangles. The exercises for Section 3 provide additional examples:

1. The relation $\perp$ (having the same parents).
2. The relation $\equiv_n$ (*congruence modulo n*) of Exercise 2.
3. The relations $\sim$, $*$, and $**$ of Exercise 3.
4. The relation R of Exercise 4.

Symbols commonly used to denote equivalence relations are $=$ (usually reserved for equality), $\equiv$, $\sim$, and $\approx$.

Suppose $\equiv$ is an equivalence relation on a set X. Each element x of X determines a certain subset of X, the set of all y in X which are $\equiv$-equivalent to x. We call this set the *equivalence class of* x and denote it by $[x]$. Formally, the definition reads: $[x] = \{y \in X \mid x \equiv y\}$.

Here are some examples:

Example 1. Let $=$ be the relation of equality on a set X. If x is any element of X, then $[x]$ is just the set consisting of x itself. $[x] = \{x\}$. Thus all equivalence classes are singletons.

Example 2. If R is the universal relation on a set X, then R is an equivalence relation and there is exactly one equivalence class: all of x. In other words, $[x] = [y]$ for all x and y in x.

Example 3. If $\perp$ is the relation "having the same biological parents" defined on H (see Exercise 1 of Section 3), then the equivalence class of $x \in H$ is precisely the set of all (biological) siblings of x together with x itself (himself? herself?).

Example 4. Consider the equivalence relation $\equiv_5$ defined in Exercise 2 of Section 3 on the set I of integers. The equivalence class of the integer x is easily seen to be the set of all integers y which can be written $y = x + 5n$, where n is an integer. Thus, for example, $[2] = \{2 + 5n \mid n \in I\} = \{\ldots, -8, -3, 2, 7, 12, \ldots\}$. A little reflection shows that many different integers determine the same equivalence class, for example, $\cdots = [-8] = [-3] = [2] = [7] = [12] = \cdots$. In fact, there are only five distinct equivalence classes in this case.

Returning to the general situation, we may ask: Given an equivalence relation $\equiv$ on a set X, what set theoretic properties does the collection of equivalence classes have?

We note, to begin with, that *each element of X belongs to some equivalence class.* For, since $\equiv$ is reflexive, $x \equiv x$ and therefore x belongs to $[x]$ for each x in x.

Second, we observe that *two equivalence classes are either the same or disjoint.* The point of this observation is that although x and y may be distinct elements, we may have $x \equiv y$ so that $[x]$ and $[y]$ have a nonempty intersection. The above statement means that if $[x] \cap [y] \neq \varnothing$, then, in fact, $[x] = [y]$. To prove it, suppose $[x]$ and $[y]$ are equivalence classes and are not disjoint. Choose $z \in [x] \cap [y]$. Then $x \equiv z$ and $y \equiv z$. Since $\equiv$ is symmetric and transitive, it follows that $x \equiv y$, whence y belongs to $[x]$. But then, by reflexivity and transitivity again, any element of $[y]$ also belongs to x, so that $[y] \subseteq [x]$. A similar argument reversing the roles of x and y shows that $[x] \subseteq [y]$.

Let us rephrase what we have proved by introducing the following concept. A collection $\mathscr{P}$ of subsets of a set X is called a *partition of X* or a *decomposition of X* provided

1. X is the union of $\mathscr{P}$ (that is, each element of X belongs to at least one member of $\mathscr{P}$).
2. Two sets in $\mathscr{P}$ are either disjoint or identical (that is, if A, B are in $\mathscr{P}$, either $A \cap B = \varnothing$ or $A = B$).

What we showed above is that *if $\equiv$ is an equivalence relation on X, then the collection $\mathscr{P}$ of equivalence classes is a partition of X.* [This partition is called the partition of X *induced by* $\equiv$ and is denoted $\mathscr{P}(\equiv)$.] It is an aesthetically satisfying and sometimes useful fact that the converse is true.

THEOREM 4.1 *Let $\mathscr{P}$ be a partition of a set X; then there is an equivalence relation $\equiv$ defined on X such that $\mathscr{P} = \mathscr{P}(\equiv)$.*

PROOF. Define a relation $\equiv$ on X as follows: $x \equiv y$ if and only if x and y belong to the same member of $\mathscr{P}$. [Formally, $\equiv$ is the subset of $X \times X$ consisting of all pairs (x, y) such that x and y belong to the same member of $\mathscr{P}$.] It is certainly clear that $\equiv$ is symmetric. Reflexivity follows from property 1 for partitions and transitivity from property 2 (Exercise 2). Thus $\equiv$ is an equivalence relation. If x is an element of X and P is the element of $\mathscr{P}$ which contains x, then P is precisely the equivalence class $[x]$ of x determined by $\equiv$. So the collection $\mathscr{P}$ is precisely the collection of all $\equiv$-equivalence classes, as asserted. ∎

From now on, then, we may specify an equivalence relation on a set either directly or by specifying a partition of X.

To illustrate this, suppose we consider the partition of the integers consisting of just two sets, the set of even integers and the set of odd integers. (These sets are disjoint and their union is the set of all integers, so they do form a partition.) Clearly the relation corresponding to this partition is the one relative to which any two even integers are equivalent (and no odd integer is equivalent to any even integer). In short, the relation is $\equiv_2$.

EXERCISES

1. Suppose $\equiv$ is an equivalence relation on a set X and x and y are elements of X. Write out in exhaustive detail the proof that if $y \in [x]$, then $[y] = [x]$, justifying each step.

2. Show in detail that $\equiv$ defined in the proof of Theorem 4.1 is an equivalence relation.

3. What partition of the real numbers is induced by the relation $x * y$ provided $\langle x \rangle = \langle y \rangle$? Here $\langle w \rangle$ denotes the largest integer less than or equal to w.

4. Fix an integer n and consider the equivalence relation $\equiv_n$ on the set of integers. If $n \neq 0$, show that there are n equivalence classes determined by $\equiv_n$. Discuss what happens when $n = 0$.

5. (a) Suppose R and S are equivalence relations on a set X. Show that $R \cap S$ is also an equivalence relation on X.

(b) Let R and S be the equivalence relations $\equiv_4$ and $\equiv_6$, respectively, defined on the set of integers. What is the intersection of these relations?

6. Let $\mathscr{S}$ be the collection of all sets. Given A and B, elements of $\mathscr{S}$, define $A \approx B$ provided there is a 1–1 function f from A onto B. Is $\approx$ an equivalence relation in $\mathscr{S}$?

7. Two relations R and S on a set X are the same (equal) provided that they are the same subsets of $X \times X$. Show that if R and S are equivalence relations on X, then $R = S$ if and only if R and S induce the same partition of X.

8. This problem is designed to show how the properties of equivalence classes change if we consider relations other than equivalence relations. Given a relation R on a set X let the *R-class* (x) of an element x of X be the set of all elements y in X such that xRy.

Thus if R is an equivalence relation, the R-class of x is just the equivalence class, $[x]$, of x.

(a) Show that if R is not reflexive, then for at least one x in X we have $x \notin (x)$.

(b) Let R be the relation defined on the set J of integers not equal to ± 1 by xRy, provided there is an integer $n \neq \pm 1$ such that x and y are multiples of n. Show that some two R-classes are not equal but have a nonempty intersection. Which of the three requirements for an equivalence relation does R fail to satisfy?

(c) Find a reflexive, transitive relation R on some subset of the real numbers such that two R-classes are not equal and not disjoint.

5. ORDER RELATIONS

A relation R in a set is called a *partial ordering* in X provided

1. R is transitive.
2. If xRy and yRx, then $x = y$.

If the domain of R is all of X we will call R a partial ordering *of* X; this will usually be the case in applications.

Note that a partial ordering is not required to be reflexive, although it may well be, and that a partial ordering is far from being symmetric; in fact, condition 2 is close to being the negation of symmetry.

We caution the reader that there is no uniformity among authors as to the definition of partial ordering. Some authors merely require condition 1; others require 1 and reflexivity; we, obviously, like the definition given above.

The standard ordering, $>$, of the real numbers is a partial ordering. Condition 2 is vacuously true since the hypothesis is never satisfied (there are no real numbers x and y such that $x > y$ and $y > x$). Note that $>$ is not reflexive; if we "make it reflexive" we get another familiar partial ordering of real numbers, $\geq$. These two orderings are read, "is strictly greater than" and "is greater than or equal to," respectively. Turning things around we get two other familiar partial orderings on the set of real numbers. These are $<$ and $\leq$, which are known to one and all as "is strictly less than" and "is less than or equal to."

The process of making a given partial ordering reflexive and of turning a partial ordering around are fairly easy to formalize. We shall do this below, but first let us look at some more examples.

Let X be a set and let $\mathscr{P}(X)$ be the collection of all subsets of X. The following are all partial orderings in $\mathscr{P}(X)$: $\subset$, $\supset$, $\subseteq$, and $\supseteq$. The domain of $\subset$ consists of all members of $\mathscr{P}(X)$ except X (why?), and the domain of $\supset$ is all of $\mathscr{P}(X)$ except $\varnothing$. The other two relations have all of $\mathscr{P}(X)$ as their domain, and so they are partial orderings of $\mathscr{P}(X)$.

In the set of integers, divisibility criteria suggest some partial orderings; for example, in Exercise 2, Section 3, we defined a relation $|$ on the set of integers by $x|y$ provided y is an integral multiple of x ($y = nx$ for some integer n). This relation is transitive but does not quite satisfy our second condition, because for every integer n, $n|-n$, and $-n|n$, but $n \neq -n$, unless $n = 0$. If we consider $|$ as a relation on I^+, the set of positive integers, then it is a partial ordering of I^+. To see this, suppose that x, y, and z are in I^+ and $x|y$ and $y|z$. Choose integers n and m such that $y = n \cdot x$ and $z = m \cdot y$; thus $z = (nm)x$, (that is, $x|z$) and we have shown that $|$ is transitive. Suppose x and y are in I^+ and $x|y$ and $y|x$. Choose integers m and n such that $x = ny$ and $y = mx$. Note that m and n must be nonnegative. Then $x = (m \cdot n) \cdot x$. It follows that $m \cdot n = 1$, and since m and n are nonnegative we must have $m = n = 1$. In particular, $x = ny = y$; this shows that condition 2 is satisfied. Additional examples are given in the exercises.

Let us formalize two simple methods for getting new partial orderings from old ones.

The first construction is that of making a partial ordering reflexive; the passage from the relation $>$ to the relation $\geq$ is a special case. Given a relation R in a set X, let $\bar{R}$ be the relation $R \cup \Delta$. (Recall that Δ is the diagonal of $X \times X$; $\Delta = \{(x, y) \in X \times X \mid x = y\}$.) In words, the new relation is defined in X by $x\bar{R}y$ if and only if xRy or $x = y$. Note that even though the domain of R might not have been all of X, the domain of $\bar{R}$ is all of X.

PROPOSITION 5.1 *If R is a partial ordering in X, then $\bar{R}$ is a reflexive partial ordering on X.*

Since $\bar{R}$ contains Δ it is certainly reflexive; the problem is to show that we have not destroyed properties 1 and 2 of a partial ordering by adding Δ to R.

Suppose then that $x\bar{R}y$ and $y\bar{R}z$ where x, y, and z are elements of X. There are a number of cases to consider. Since $x\bar{R}y$ we have (a) ${}_xR_y$ or (b) $x = y$. Since $y\bar{R}z$ we have (c) yRz or (d) $y = z$. If (a) and (c) hold we have, by transitivity of R, xRz and therefore $x\bar{R}z$. If (a) and (d) or (b) and (c) hold, then by direct substitution we get xRz and therefore $x\bar{R}z$. If (b) and (d) hold, then $x = y = z$, so again $x\bar{R}z$, whence $\bar{R}$ is transitive. We leave to the reader the verification of condition 2 for a partial ordering.

The notion of "reversing" a partial ordering is formalized as follows. If R is a relation in a set X, then we let $\hat{R}$ be the relation given by $\hat{R} = \{(y, x) \in X \times X \mid (x, y) \in R\}$. In other words, we have $y\hat{R}x$ if and only if xRy.

PROPOSITION 5.2 *If R is a partial ordering in X, then $\hat{R}$ is a partial ordering in X. If R is reflexive, so is $\hat{R}$.*

The proof is easy and we omit writing it out. In general, the domain of R and $\hat{R}$ need not be the same.

Let us adopt the following standard convention. From now on a partial ordering on an abstract set X will be denoted by $>$. In this case $\geq$ denotes the partial ordering obtained by adding the diagonal to $>$ (that is, what we get by making $>$ reflexive); also, $<$ and $\leq$ will denote the partial orderings obtained by reversing $>$ and $\geq$.

EXERCISES

1. Show that the relation $\bar{R}$ of Proposition 5.1 satisfies the second condition for a partial ordering.

2. Prove Proposition 5.2.

3. Give an example of a partial ordering R in a set such that the domains of R and $\hat{R}$ (defined above) are different. (There is a simple, natural example of this.) Give an example with the same punch line, where R is a partial ordering *on* X (that is, where the domain of R is all of X).

4. Let $\mathscr{F}$ be the collection of functions mapping R, the set of real numbers, into itself. Which of the following are partial orderings on $\mathscr{F}$?

(a) $f > g$, provided $f(x) > g(x)$ for all $x \in R$.
(b) $f > g$, provided that for some real number t, $f(x) > g(x)$ for all $x > t$.
(c) $f > g$, provided $f(x) + g(x) > 0$ for all x.

5. Let X be a set and let $>$ be a partial ordering on X. Denote by $\mathscr{P}(X)$ the collection of all subsets of X and define a relation $\vdash$ on $\mathscr{P}(X)$ by $A \vdash B$ provided that for $x \in A$ and $y \in B$, we have $x \geq y$. Show that $\vdash$ is a partial ordering in $\mathscr{P}(X)$. How about the relation $\rightarrow$ defined on $\mathscr{P}(X)$ by $A \rightarrow B$ provided that for some $x \in A$ and all $y \in B$, $x \geq y$?

6. Let $>$ be a partial ordering on a set X and let $\geq$ have the meaning assigned by the standing convention. Show that if $x \geq y$ and $y \geq x$ in X, then $x = y$.

6. ADDITIONAL PROPERTIES OF ORDER RELATIONS

The object of this section is to formalize some notions with which the reader is probably already familiar. Roughly speaking we are going to look at maxima and minima in a partially ordered set.

Throughout this section, $>$ denotes some fixed partial ordering on a set X. The symbols $\geq$, $<$, and $\leq$ denote the related partial orderings defined at the end of Section 5.

Let A be a subset of X and let x be an element of X. We say that x is an *upper bound* for A in case $x \geq a$ for every element a of A. Similarly, x is a lower bound for A if $x \leq a$ for every a in A.

We say that x is a *least upper bound* for A if the following two conditions hold:

1. x is an upper bound for A.
2. If y is any upper bound for A, then $x \leq y$.

These two conditions can be stated in another way. Let $U(A)$ be the set of all upper bounds for A in X. Then x is a least upper bound for A provided

1. x is an element of $U(A)$.
2. x is a lower bound for $U(A)$.

Here are several illustrations of these concepts. Let $>$ be the usual partial ordering of the set R of real numbers. Let x_0 be a fixed real number and let A be the set of real numbers x such that $x < x_0$. A number y is an upper bound for A if and only if $y \geq x_0$; in the terminology introduced just above, $U(A) = \{y \in R \mid y \geq x_0\}$. The set A has a least upper bound in R, namely, x_0.

Suppose we change things a bit by considering $>$ as a partial ordering of the subset X of R consisting of all real numbers except x_0. Let A be as before; then A is a subset of x and a number y in X is an upper bound for A if and only if $y > x_0$. In this new setting (that is, in X) A has no least upper bound. For suppose $y \in X$ is any upper bound for A. Since x_0 is not in X we have $y > x_0$. From the usual ordering properties of real numbers it follows that the number $(y + x_0)/2$ is an upper bound for A and is strictly less than y. Hence, no upper bound for A is a least upper bound.

Again let R be the set of real numbers and consider the following subsets of R; R itself, $\varnothing$, $A = \{x \in R \mid x \text{ is an integer}\}$, $B = \{x \in R \mid x = 1 - (1/n) \text{ for some positive integer } n\}$. The first of these, R itself, has no upper bounds and, a fortiori, no least upper bounds. The empty set has plenty of upper bounds in R; in fact, every real number is an upper bound for $\varnothing$. The subset A of R has no upper bound in R. The set B has upper bounds in R and, in fact, it has a least upper bound, the number 1.

The astute reader will have noticed in those cases where a subset of R has a least upper bound, this least upper bound is unique. In other words, a subset of R cannot have two distinct least upper bounds. This is true in any partially ordered set.

PROPOSITION 6.1 *Let $>$ be a partial ordering on a set X and let A be a subset of X. Suppose x and y are least upper bounds for A in X; then $x = y$.*

PROOF. The two conditions on least upper bounds yield the inequalities $x \geq y$ and $y \geq x$. By Exercise 6 of Section 5, $x = y$. ▮

Let us look at some slightly less familiar orderings.

Earlier we defined a partial ordering $|$ on the set I^+ of positive integers by $x \mid y$ provided $y = nx$ for some n in I^+. Thus $x \mid y$ is shorthand for "y is a multiple of x" or "x divides y." Let A be the set consisting of the integers 2, 3, and 5. An element y of I^+

is an upper bound for A if and only if it is divisible by 2, 3, and 5. Thus 30 and 60 are upper bounds for A but 6, 15, and 10 are not.

To discuss least upper bounds we need the following result from number theory: *If x, y, and z are positive integers, z is prime and z divides xy, then z divides x or z divides y.* This is one of the fundamental results in elementary number theory. (A proof can be found in N. H. McCoy, *Introduction to Modern Algebra*, Allyn & Bacon, Inc., Boston, 1967, p. 59.) Using this result we can show that the least upper bound of A is the integer 30. We know that 30 is an upper bound; we need show that 30 divides every upper bound. If x is an upper bound, then x is divisible by 2, say $x = 2y$ for some y in I^+. Now x is divisible by 3, which is prime, and 3 does not divide 2. By the result quoted above, 3 must divide y, say $y = 3z$. Substituting we see that $x = 2(3z) = 6z$. Now 5 divides x, but 5 does not divide 6, hence 5 divides z, say $z = 5w$. Thus, finally, we have $x = 6(5w) = 30w$, for some w in I^+: that is, 30 divides x. This is what we wanted to prove.

As a final example, let X be a set and let $\mathscr{P}(X)$ be the collection of all subsets of X, partially ordered by $\subseteq$. Let $\mathscr{A}$ be a subcollection of $\mathscr{P}(X)$; that is, $\mathscr{A}$ is a collection of subsets of X. A set B in $\mathscr{P}(X)$ is an upper bound for $\mathscr{A}$ if and only if $A \subseteq B$ for every A in $\mathscr{A}$. It should be crystal clear that the least upper bound of $\mathscr{A}$ is the set $\cup\mathscr{A}$.

At the same time we defined the term upper bound, we defined the "dual" concept of lower bound. The concept of least upper bound has a dual, greatest lower bound. Precisely, if A is a subset of a set X, then an element x of X is a *greatest lower bound* of A provided

1. x is a lower bound for A.
2. If y is any lower bound for A, then $y \leq x$.

Thus the concepts of lower bound and greatest lower bound are obtained by reversing the inequalities (that is, replacing $\geq$ by $\leq$) in the definition of upper bound and least upper bound. In general, every definition and proposition we state in terms of a partial ordering $\geq$ has a dual statement obtained by reversing inequalities. It may turn out, as in the definition of linear ordering given in one of the exercises below, that a statement and its dual say the same thing; however, usually the two statements are distinct assertions.

If one reverses inequalities in the proof of a given proposition, one obtains the proof of the dual proposition. For example, by reversing inequalities in the Proposition 6.1, one obtains a proof of the uniqueness of greatest lower bounds (see Exercise 1).

The exercises given below are designed to give the reader practice in manipulating upper and lower bounds. In addition, one or two concepts involving order relations are introduced. Thus the exercises should be considered an integral part of the text of this section.

EXERCISES

1. Prove the following statement, which is the dual of Proposition 6.1: *Let $>$ be a partial ordering on a set X and let A be a subset of X. If x and y are greatest lower bounds for A in X, then $x = y$.*

2. Let $>$ be a partial ordering on a set X and let $<$ be the dual partial ordering. Let A be a subset of X and let x be an element of X. Show that

   (a) x is an upper bound for A relative to $>$ if and only if x is a lower bound for A relative to $<$.

   (b) Formulate and prove corresponding statement for least upper and greatest lower bounds.

3. Let $>$ be a partial ordering on a set X. Given a subset Y of X, let $U(Y)$ and $L(Y)$ denote the set of upper and lower bounds of Y, respectively. Let A be a subset of X.

   (a) Show that $A \subseteq U(L(A))$, which says that every element of A is an upper bound of the set of all lower bounds for A.

   (b) Show that $A \subseteq L(U(A))$; explain in words what this means.

   (c) Show that $L(A) \subseteq L(U(A))$. Explain in words what this means.

   (d) State and prove the dual of part (c).

4. Let $>$ be a partial ordering on a set X and suppose X has the property that if A is a nonempty subset of X and A has an upper bound, then it has a least upper bound. Prove that the dual statement is also true. That is, prove that if B is a nonempty set which has a lower bound, then B has a greatest lower bound. (*Hint*: Given B as above, let A be the set of lower bounds of B. Does A have an upper bound?)

5. Let $>$ be a partial ordering on a set X. We say X is complete (relative to $>$) if X satisfies one (and hence both) of the conditions stated in Exercise 4.

(a) Let $\subseteq$ be the usual partial ordering of $\mathscr{P}(X)$, where X is some set. Is $\mathscr{F}(X)$ complete with respect to $\subseteq$?

(b) Assume the set R of real numbers is complete with respect to $\geq$. Let a and b be two real numbers, with $a < b$. Show that the closed interval from a to b, $[a, b] = \{x \in R \mid a \leq x \leq b\}$, is also complete with respect to $\geq$.

6. A partial ordering $\geq$ of a set X is called a *linear ordering* of X in case every two elements of X are comparable with respect to $>$; that is, if x and y are elements of X, then $x \geq y$ or $x \leq y$. Which of the following are linear orderings and which are not?

(a) The usual ordering $\geq$ on the set of real numbers.

(b) Set containment $\subseteq$ on $\mathscr{P}(X)$, where X is a set with at least two elements.

(c) The relation $\mid$ defined earlier in Exercise 2, Section 2, on the set I^+ of positive integers.

CHAPTER 4

Cardinality

1. INTRODUCTION

In this chapter we assume a knowledge of the basic properties of the integers. We shall use the principle of induction in various forms and we shall use the following special case of the unique factorization theorem: If x is an integer and if there exist integers a, b, a', b' such that $x = 2^a 3^b = 2^{a'} 3^{b'}$, then $a = a'$ and $b = b'$.

2. FINITE SETS

We all have an intuitive idea of what it means to say, "There are exactly 25 people in this room" or "The set $\{a, b, c\}$ has exactly eight different subsets." In general, the verification of such statements can be reduced to the problem of counting the elements of a set.

The way one counts a set is to assign, verbally or otherwise, the integer 1 to one element of the set, the integer 2 to another element of the set, and to continue in this way. No integer is skipped or assigned to two different elements, and every element of the set has some integer assigned to it. The last integer named is the number of elements in the set.

We now formalize the counting process. Let n be a positive integer; we say that a set *X has n elements*, or that the *number of elements in X is n*, provided there is a 1–1 function from X onto the set $\{1, 2, \ldots, n\}$. (Note that since 1–1 onto functions have inverses, a set X has n elements if and only if there is a 1–1 function from $\{1, 2, \ldots, n\}$ onto X.) Naturally, we also say that the empty set has zero elements.

In general, we shall say that a set is *finite* if it has n elements, for some nonnegative integer n.

The definition of "finite" is more powerful than it might seem at first glance. For if we are given a nonempty finite set X then, for free, we are also given the existence of a function which "counts" X, that is, a 1–1 function from X onto a set of the form $\{1, 2, \ldots, n\}$. In general one solves a problem involving combinatorial properties of finite sets by translating it, via "counting functions," to a problem about integers and solving this problem using structural properties of the integers.

These remarks will be illustrated by the proofs which follow. To begin with we prove a useful fact about integers. As a space-saving device, let us use the symbol A_n to denote the set $\{1, 2, \ldots, n\}$.

LEMMA 2.1 *For each positive integer n, every subset of A_n is finite.*

PROOF. For each positive integer n, let $S(n)$ be the assertion that every subset of A_n is finite. Certainly $S(1)$ is true.

Suppose that $S(n-1)$ is true where $n \geq 2$ is an integer; we will show that $S(n)$ is also true. Let M be a subset of A_n. In case $M = \varnothing$ or $M = A_n$, M is finite (why?) and we are done.

The remaining possibility is that M is a proper nonempty subset of A_n. In this case there is some element m of A_n such that $M \subseteq A_n - \{m\}$. Define $g: A_n - \{m\} \to A_{n-1}$ by

$$g(x) = \begin{cases} x & \text{if } x < m; \\ x - 1 & \text{if } x > m \end{cases}$$

then g is 1–1 and onto. (The reader should verify this unless he has already done Exercise 5, Section 3, Chapter 2.)

Now $g(M)$ is a subset of A_{n-1} and, since $S(n-1)$ is true, $g(M)$ is finite. Since $g(M) \neq \varnothing$, there is a positive integer m and a 1–1 function f of $g(M)$ onto A_m. The composition $f \circ g$ is a 1–1 function from M onto A_m, so M is finite.

To summarize, we know that $S(1)$ is true and that, if $S(n-1)$ is true, then so is $S(n)$. By induction, $S(n)$ is true for every positive integer and we have proved the lemma. ∎

The next theorem states three important facts about finite sets. We have stated it so as not to be bothered by the empty set, which is finite by definition.

THEOREM 2.2 **Theorem on Finite Sets**

(a) *A nonempty subset of a finite set is finite.*
(b) *If A and B are finite nonempty sets, then $A \times B$ is finite.*
(c) *If $\mathscr{F}$ is a finite nonempty collection of finite nonempty sets, then $\cup \mathscr{F}$ is finite.*

PROOF.

(a) Suppose $B \neq \varnothing$ is a subset of the finite set A. Certainly

$A \neq \varnothing$, so there is a 1–1 function f from A onto A_n, where n is a positive integer. By the lemma, $f(B)$, being a subset of A_n, is finite and there is a 1–1 function g from $f(B)$ onto A_m for some positive integer m. Then $g \circ f$ is 1–1 from B onto A_m; hence B is finite.

(b) Suppose A has m elements and B has n elements, where m and n are positive integers. Choose 1–1 functions f and g from A and B onto A_m and A_n, respectively. Define h: $A \times B \to I^+$ (the set of positive integers) as follows:

$$h((a, b)) = 2^{f(a)}3^{g(1)}.$$

To see that h is 1–1, suppose $h((a, b)) = h((a', b'))$, that is, $2^{f(a)}3^{g(b)} = 2^{f(a')}3^{g(b')}$. By the unique factorization theorem we must have $f(a) = f(a')$ and $g(b) = g(b')$. Since f and g are 1–1, this implies that $a = a'$ and $b = b'$; hence $(a, b) = (a', b')$.

Now for every (a, b) in $A \times B$ we have $6 \leq h(a, b) \leq 2^m 3^n$, so $h(A \times B)$ is a subset of $A_{2^m 3^n}$ and thus $h(A \times B)$ is finite. Since h is 1–1 it follows as usual that $A \times B$ is finite.

(c) Let $\mathscr{F}$ be a nonempty finite collection of nonempty finite sets. Choose a positive integer m and a 1–1 function f from A_m onto $\mathscr{F}$. Now for each integer i in A_m, let n_i be the number of elements in the set $f(i) \in \mathscr{F}$. Let g_i be a 1–1 function from $f(i)$ onto A_{n_i}.

Let N be the largest of the integers $n_1, n_2, \ldots, n_m$. Note that each g_i is a function from $f(i)$ into A_N.

Now define h: $\cup \mathscr{F} \to A_m \times A_N$ by $h(x) = (i, g_i(x))$, where i is the least integer such that $f(i)$ contains x.

The function h is 1–1, for suppose $h(x) = (i, g_i(x))$ is equal to $h(y) = (j, g_j(y))$. By the definition of equality of ordered pairs, $i = j$; hence $g_i = g_j$ and, since g_i is 1–1 and $g_i(x) = g_i(y)$, we must have $x = y$.

The set $A_m \times A_N$ is finite, by part (b); therefore, $h(\cup \mathscr{F})$ is finite, by part (a), and, since h is 1–1, it follows that $\cup \mathscr{F}$ itself is finite. This completes the proof of part (c). ∎

EXERCISES

1. Prove that if f: $X \to Y$ is 1–1 and $f(X)$ is finite, then X is finite. (This is a formalization of a fact we used three times in proving the theorem on finite sets.)

2. Let $f: X \to Y$ be a function with the property that, for each y in Y, the set $\{x \in X \mid f(x) = y\}$ is finite. Show that if Y is finite, then so is X. (Use the theorem on finite sets.)

3. (a) Let $S(n)$ be the statement that there is no 1–1 function from A_n onto a proper subset of itself. Prove that $S(n)$ is true for every positive integer n. (Use Exercise 5, Section 3, Chapter 2, to carry out the induction step.)
(b) Prove that there is no 1–1 function from a finite set onto a proper subset of itself.

4. Show that if, according to our definition, a set has n elements and also has m elements, then $m = n$. (Use Exercise 3.)

5. A subset A of a set X is called *cofinite* if X-A is finite. Show that if $\mathscr{A}$ is any collection of cofinite subsets of a set X then $\cup\mathscr{A}$ is cofinite. Show that if $\mathscr{A}$ is a finite collection of cofinite subsets $\cap\mathscr{A}$ is cofinite. Give an example showing that we cannot delete the word "finite" in the preceding sentence.

3. INFINITE SETS

A set which is not finite is called *infinite*. Now the statement that X is not finite is precisely the statement that X is not empty, and, moreover, for every positive integer n there is no 1–1 function from X onto A_n. Thus it would seem that to prove a given set is infinite, one would have to prove the nonexistence of functions of a certain type. This formidable-looking problem can be circumvented by finding conditions which are equivalent to being infinite but which are easier to verify.

THEOREM 3.1 **Characterization of Infinite Sets** *Let X be a set; the following are equivalent:*

(a) *There is a* 1–1 *function from X onto a proper subset of X.*
(b) *X is infinite.*
(c) *There is a* 1–1 *function from I^+ into X.*

PROOF. We shall prove the following chain of implications: (a) $\Rightarrow$ (b) $\Rightarrow$ (c) $\Rightarrow$ (a).

The implication (a) $\Rightarrow$ (b) is merely the contrapositive of the result stated in Exercise 3 (b), Section 2.

To prove that (b) $\Rightarrow$ (c), suppose X is an infinite set. We define a function f from I^+ into X inductively as follows: Let $f(1)$ be any element of X and suppose $f(1), f(2), \ldots, f(n)$ have been defined and are distinct elements of X. Since X is not finite, the set $A = \{f(1), \ldots, f(n)\}$ is a proper subset of X. Choose $x \in X - A$ and define $f(n+1) = x$.

By induction $f(n)$ is defined for every $n \in I^+$. By the construction of f, if $n \neq m$, then $f(n) \neq f(m)$; thus f is 1–1.

To prove that (c) $\Rightarrow$ (a), suppose we have a set X and a 1–1 function $f\colon I^+ \to X$. Let $g\colon I^+ \to I^+$ be the function defined by $g(x) = 2x$; note that g is 1–1. Now define $h\colon X \to X$ by the formula

$$h(x) = \begin{cases} x & \text{if } x \notin f(I^+), \\ f \circ g \circ f^{-1}(x) & \text{if } x \in f(I^+). \end{cases}$$

This function takes X into itself. To see that it is not onto X (hence is onto a proper subset of X) let us consider the element $f(1) \in X$. We assert that $f(1)$ is not the image under h of any element of X. If $x \notin f(I^+)$, then $h(x) = x$, so $h(x) \neq f(1)$. If $x \in f(I^+)$, then there is a unique integer n such that $x = f(n)$. Then $h(x) = f \circ g \circ f^{-1}(x) = f \circ g(n) = f(2n) \neq f(1)$. So $f(1)$ is not in $h(X)$; thus $h(X)$ is a proper subset of X. ∎

Using this theorem the reader should check that the following sets are infinite: I, I^+ (the set of positive integers), $\{x \in I \mid x \text{ is divisible by } 5\}$, the set of rational numbers, the set of real numbers.

In some sense I^+ serves as a measuring device to differentiate between finite sets and infinite sets; a set is infinite if and only if it contains a "copy" of I^+ (that is, a subset which is the range of a 1–1 function defined on I^+) and finite otherwise.

It is useful to give a special name to those infinite sets which are "copies" of I^+. We shall say that a set X is *countably infinite* provided there is a 1–1 function from X onto I^+. Observe that since 1–1 onto functions have inverses, we could just as well have required that the function be 1–1 from I^+ onto X.

It is easily seen that I^+, I, and $\{x \in I \mid x \text{ is divisible by } 5\}$ are countably infinite. Often we do not need to distinguish between finite and countably infinite sets, and we shall say that a set is *countable* provided it is finite or countably infinite.

We will now prove some results analogous to those of the

preceding section. First we note that *every subset of* I^+ *is countable*. To prove this it suffices to show that every infinite subset of I^+ is countably infinite. Thus given an infinite subset X of I^+ we wish to construct a 1–1 function from I^+ onto X. Such a function may be defined inductively by defining $f(1)$ to be the least element of X and, having defined $f(1), \ldots, f(n)$, by defining $f(n+1)$ to be the least element of $X - \{f(1), \ldots, f(n)\}$. The function defined this way clearly has the desired properties.

The following result will be used in the proof of our next theorem.

THEOREM 3.2 **Alternative characterization of Countability**
A nonempty set X is countable if and only if there is a 1–1 *function from X into I^+.*

PROOF. If X is countable, it is either finite or countably infinite. In the first case apply the definition of finite set. In the second case apply the remark following the definition of countably infinite set. In either case, a function with the desired properties is obtained.

Conversely, suppose f: $X \to I^+$ is 1–1. Now $f(X)$ is a subset of I^+ and hence is countable. If $f(X)$ is finite, so is X. If $f(X)$ is countably infinite let $g: f(X) \to I^+$ be a 1–1 onto function and note that $g \circ f$: $X \to I^+$ is 1–1 and onto, whence X is countably infinite. ▮

THEOREM 3.3 **Theorem on Countable Sets**

(a) *A subset of a countable set is countable.*

(b) *If A and B are nonempty countable sets, then $A \times B$ is countable.*

(c) *If $\mathscr{F}$ is a nonempty countable collection of nonempty countable sets, then $\cup \mathscr{F}$ is countable.*

PROOF.

(a) Suppose $B \subseteq A$, where A is countable. If $B = \varnothing$, then it is finite, hence countable. If $B \neq \varnothing$, then A is a nonempty countable set so there is a 1–1 function f from A into I^+. Let g be the restriction of f to B; that is, g is the function defined on B by $g(x) = f(x)$. Then g: $B \to I^+$ is 1–1 so, by the preceding result, B is countable.

(b) If A and B are nonempty and countable, let $f\colon A \to I^+$ and $g\colon B \to I^+$ be 1–1 functions. Then $h\colon A \times B \to I^+$ defined by $h((a, b)) = 2^{f(a)}3^{g(b)}$ is 1-1, so $A \times B$ is countable.

(c) We leave the proof of this part as Exercise 1. ▌

The proof of part (b) of this theorem has the advantage of being rigorous and easy to understand once one has seen the proof of part (b) of the theorem on finite sets. It has the disadvantage of exhibiting a rather unnatural way to count $A \times B$. We pause here to sketch another proof of part (b); this is the usual proof and can be formalized, although we will not set up the machinery to do so.

Let us suppose we have two countable sets A and B which, for convenience, we will assume to be infinite. Choose functions $a\colon I^+ \to A$, $b\colon I^+ \to B$ which are 1–1 and onto. Thus a and b are sequences and we write the elements of A as $a_1, a_2, \ldots,$ where $a_i = a(i)$ for each i in I^+. Similarly, we write the elements of B as $b_1, b_2, \ldots,$ where $b_i = b(i)$. The product $A \times B$ can be visualized as shown in Figure 4.1.

b_j	(a_1, b_j)					(a_i, b_j)
b_3	(a_1, b_3)					
b_2	(a_1, b_2)	(a_2, b_2)				
b_1	(a_1, b_1)	(a_2, b_1)	(a_3, b_1)			(a_i, b_1)
	a_1	a_2	a_3	a_4		a_i

FIGURE 4.1

We now indicate how to define a 1–1 function from I^+ onto $A \times B$. The first few values of the functions are $f(1) = (a_1, b_1)$, $f(2) = (a_2, b_1)$, $f(3) = (a_1, b_2)$, $f(4) = (a_1, b_3)$, $f(5) = (a_2, b_2)$

and $f(6) = (a_3, b_1)$. Continue defining f by making it weave diagonally back and forth as shown in Figure 4.2.

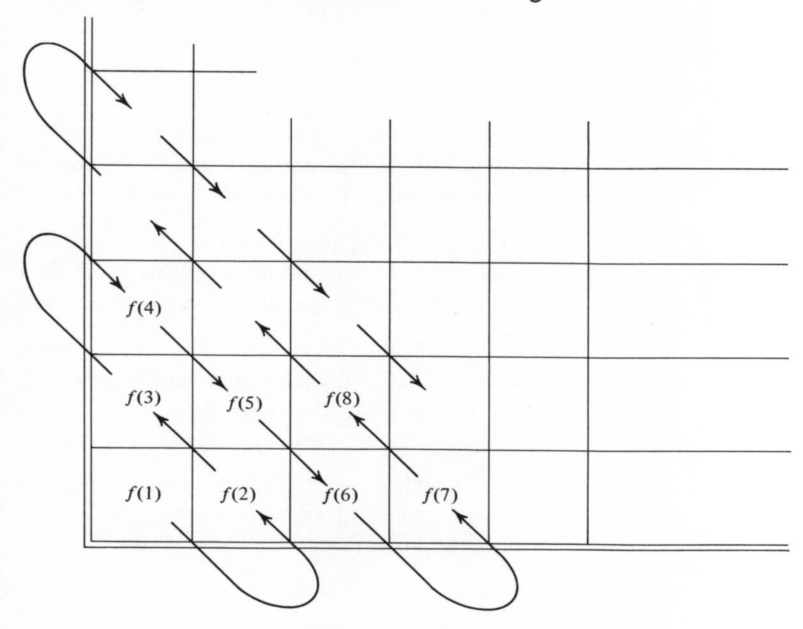

FIGURE 4.2

Thus $f(i)$ is the element of $A \times B$ which occurs in the ith box "encountered" by f. Intuitively, this process defines a function with the desired properties.

A similar trick can be used to prove part (c) directly.

EXERCISES

1. Prove part (c) of the theorem on infinite sets by using part (b) of that theorem. Sketch an alternative proof based on rectangular arrays and a "sweep-out" function.

2. Prove or disprove the following statements:
(a) If $\mathscr{F}$ is a family of countable sets, then $\cap\mathscr{F}$ is countable.
(b) If $\mathscr{F}$ is a family of countably infinite sets, then $\cap\mathscr{F}$ is countably infinite.

3. Suppose $f\colon X \to Y$ has the property that, for each y in Y, the set $\{x \in X \mid f(x) = y\}$ is countable. Show that if Y is countable, then X is also countable.

4. Let C be the collection of all functions from I^+ into the set consisting of the two numbers -1 and 1. Prove that C is infinite by exhibiting a 1–1 function from I^+ into C. Can you prove that C is countable?

5. Prove the following:

(a) If X is any nonempty set, then there is a sequence in X.
(b) A set is countably infinite if and only if it can be written as (the range of) a sequence no two terms of which are equal.

6. A subset A of a set X is called *cocountable* if $X - A$ is countable. Show that if $\mathscr{A}$ is a collection of cocountable subsets of X then $\cup\mathscr{A}$ is cocountable. Formulate and prove the corresponding result for $\cap\mathscr{A}$.

7. Let $\{A_n\}$ be a sequence of subsets of a set X and suppose each A_n is infinite. Show how to obtain a sequence $\{x_n\}$ in X, no two terms of which are equal, such that x_n belongs to A_n.

8. Show that the set of all polynomials with integer coefficients is countable. One might begin by considering, for a fixed nonnegative integer n, the set of polynomials of degree n.

4. UNCOUNTABLE SETS

Although countable sets are nice to work with, because they can be investigated using properties of the integers, not all the sets one meets in mathematics are countable. In fact, because of some theorems from a branch of mathematics known as topology, it can be shown that if one restricts his attention to countable sets only, then many (if not most) of the results of classical analysis, including calculus, have to be discarded. On a more immediate level for readers of this book is the fact the set of real numbers happens not to be countable.

A set which is not countable is called *uncountable*. We are not in a position yet to prove that the set of real numbers is uncountable. This will be postponed until we have investigated the real numbers more thoroughly.

We content ourselves with giving one simple example of a set which is uncountable and presenting some observations about such sets.

Let F be the set of all functions from I^+ into the set consisting of the two numbers -1 and 1. We assume that F is countable and derive a contradiction to one of our previous results. First we observe that F is not finite (this is Exercise 4 of Section 3). Thus F is countably infinite and there is a function $f: I^+ \to F$ which is 1–1 and onto. (We will not use the 1–1-ness of f in what follows.)

Define $g: I^+ \to \{-1, 1\}$ by the formula $g(n) = -(f(n))(n)$. [For example, suppose we want to find out what $g(5)$ is. We first look at the image of 5 under f; this is one of the functions in F; that is, $f(5)$ is a function from I^+ into $\{-1, 1\}$. If its value at 5 is 1, then the value of g at 5 is -1, and vice versa.]

Now g is in C, hence it is the image under f of some integer m, i.e., $g = f(m)$. But this is impossible because, by definition of g, g and $f(m)$ have different values at m. So, after all, no such function f exists and C is not countable.

A proof very much like this one will be used later to show that the reals are not countable. Roughly speaking, using decimal expansions, we will identify the set of reals between 0 and 1 with a subset S of the set of functions from I^+ into the set $\{0, 1, \ldots, 9\}$ and then show that there is no function from I^+ onto S.

EXERCISES

1. Show that if X is a countable subset of an uncountable set Y, then $Y - X$ is uncountable.

2. Show that if $f: X \to Y$ is onto, Y is countable and X is uncountable, then there is some $y \in Y$ such that $\{x \in X \mid f(x) = y\}$ is uncountable.

3. Associated with each subset X of I^+ let $f_X: I^+ \to \{-1, 1\}$ be the function defined by: $f_X(x) = \begin{cases} 1 & \text{if } x \in X, \\ -1 & \text{if } x \notin X. \end{cases}$ Thus, for example, $f_\varnothing$ is identically -1 and f_{I^+} is identically 1.

Let G be the collection of all such functions; $G = \{f_X \mid X \subseteq I^+\}$.

(a) Show that there is a 1–1 function from $\mathscr{P}(I^+)$, the collection of all subsets of I^+, onto G.

(b) Show that there is a 1–1 function from G onto the set C of this section.

(c) Putting (a) and (b) together, what do you conclude about $\mathscr{P}(I^+)$?

4. Farmer Brown runs a large chicken ranch having a countably infinite set of nests. It is a matter of fact that if every chicken occupies one nest, then at least two chickens must occupy the same nest. Prove that when every chicken occupies one nest some nest is occupied by uncountably many chickens.

5. Let X be countable set and Y an uncountable set. Show that there exist functions $f: X \to Y$ and $g: Y \to X$ such that $g \circ f$ is the identity on X. Do there exist functions $f: X \to Y$ and $g: Y \to X$ such that $f \circ g$ is the identity on Y?

5. CARDINALITY

This section has a twofold purpose. First, we wish to introduce the concept of cardinality which to some extent unifies the preceding three sections and which justifies the title of this chapter. Second, we shall try to give the reader some notion of the ideas one works with in studying cardinality.

The reader will notice that the notions "finite," "infinite," "countable," and "uncountable" all depend on the idea of comparing sets by means of 1–1 functions. Let us say that a set X has the same *cardinality* as a set Y, and write $X \approx Y$, provided there is a 1–1 function from X onto Y. Using elementary facts about 1–1 functions, we then have the following rules (where X, Y, and Z are arbitrary sets):

5.1. $X \approx X$.

5.2. If $X \approx Y$, then $Y \approx X$.

5.3. If $X \approx Y$ and $Y \approx Z$, then $X \approx Z$.

Thus $\approx$ is an equivalence relation.

Here, in disguise, are some definitions and results of the preceding sections which the reader might wish to identify.

1. X is finite if and only if $X = \varnothing$ or, for some n, $X \approx A_n$.
2. If $A \approx I^+$ and $B \approx I^+$, then $A \times B \approx I^+$.
3. X is countable if and only if there is a subset Y of I^+ such that $X \approx Y$.
4. $I^+ \not\approx \{f | f: I^+ \to \{-1, 1\}\}$. (Slash means negation.)
5. $\mathscr{P}(I^+) \approx \{f | f: I^+ \to \{-1, 1\}\}$.

Generalizing this notion somewhat, let us define a relation $\preccurlyeq$ between sets as follows: $X \preccurlyeq Y$ provided there is a 1–1 function from X into Y. If $X \preccurlyeq Y$ but $X \not\approx Y$, we write $X \prec Y$.

The statement that the set R of real numbers is uncountable becomes $I^+ \preccurlyeq R$. The order relation $\preccurlyeq$ behaves somewhat like the standard ordering between numbers. In particular we have the following rules, where X, Y, and Z are any sets:

5.4. $X \preccurlyeq X$.
5.5. If $X \preccurlyeq Y$ and $Y \preccurlyeq Z$, then $X \preccurlyeq Z$.
5.6. If $X \preccurlyeq Y$ and $Y \preccurlyeq X$, then $X \approx Y$.

The first two statements are completely trivial to prove but the last is hard. It is called the Schroeder–Bernstein theorem. Some idea of its difficulty may be gained by writing out what it says in terms of functions.

It is possible to assign to any set X an object called the *cardinal number of* X. This is not a number in the usual sense, although in the case of a finite set the cardinal number behaves like the number of elements in the set. There is even an arithmetic of cardinal numbers; roughly speaking, sums of cardinal numbers correspond to unions of sets and products to Cartesian products. For example, the cardinal number of I^+ is denoted $\aleph_0$ and in cardinal arithmetic the product of $\aleph_0$ with itself is $\aleph_0$; this corresponds to the fact that if A and B are countably infinite, then so is $A \times B$.

EXERCISES

1. Prove Statements 5.1, 5.2, and 5.3 using results about 1–1 onto functions.

2. Relate each of statements 1 through 5 of this section to a result or definition of a previous section.

3. Verify Statements 5.4 and 5.5. Write out Statement 5.6 in terms of functions.

4. Let us say that a real number x is of type a if there is a polynomial function p, not identically zero, with integer coefficients such that $p(x) = 0$. Let A be the set of all real numbers of type a. Is A finite, countably infinite, or uncountable? You should recall that if p is a nonzero polynomial of degree k the equation $p(x) = 0$ has at most k roots.

5. Assuming there exists a 1–1 map from $R \times R$ onto R show that $R \approx R \times R$ and that $R \approx R \times R \times R$ (the set of all ordered triples of real numbers). What is the most general result along these lines that you can prove?

CHAPTER 5

The Real Numbers

1. INTRODUCTION

This chapter is devoted to studying the fundamental properties of the basic system in mathematics, the real number system. The reason such studies are useful is that the real numbers carry an algebraic structure, an order structure, and a distance structure. These combine with each other to provide a system which is simple enough to serve as a training ground for beginning mathematicians and yet rich enough to yield useful applications to the real world.

The area of mathematics known as analysis, which includes calculus, has grown out of the study of functions whose domain and range are sets of real numbers. Even the more exotic systems which are the object of current mathematical research are, for the most part, built up from the system of real numbers or modeled after it in some way.

Logically the first step in our study should be to say explicitly what the real number system is. One possible way to do this is to start with some simpler system which we know (or with some axioms) and then to construct the real numbers. The properties of the system can then be deduced by analyzing the construction process. We shall use a different method of approach. We begin by stating what properties the system is to have and then sketch a way of constructing such a system. The construction process is time consuming and, at one point, difficult; for these reasons we feel that the details should be filled in later. We end the section with some geometric considerations which justify the standard picture of the real numbers as points on a line.

The real number system can completely and briefly be described as an ordered field which satisfies the Dedekind property. Let us examine these requirements one by one. From now on we use R to denote the set of real numbers.

First, R is a field. This says that there are two operations, called addition and multiplication, which behave according to certain abstract rules. Rather than stating these rules here we merely note that these turn out to be the usual operations, $+$ and $\cdot$, with the usual properties. (The axioms for a field are given in Appendix 1.)

In particular, if a, b, and c are real numbers with $a \neq 0$, then the equation $a \cdot x + b = c$ has a solution x in R.

Second, there is a partial ordering, denoted $<$, defined on R which satisfies these requirements:

1. If x is in R, then exactly one of the following holds: $x=0$, $x<0$, $0<x$.
2. If $0<x$ and $0<y$, then $0<x+y$.
3. If $0<x$ and $0<y$, then $0<x\cdot y$.

An element x of R is *positive* provided $0<x$; the terms negative, nonpositive, and nonnegative also have their usual meanings.

Finally, R has what we call

THE DEDEKIND PROPERTY. *Suppose A and B are nonempty subsets of R such that $R=A\cup B$ and for every a in A and b in B we have $a<b$. Then there is an element $c\in R$ such that for every a in A and every b in B, $a\leq c$ and $c\leq b$.*

Here is a brief description of how one constructs a system with these properties. As a starting point we assume the existence of the set I of integers (positive, negative, and zero) with the usual addition, multiplication, and ordering. In one sense, the integers possess a rather primitive algebraic structure; for, although any equation of the form $x+a=b$, with a, b in I, has a solution x in I, very few equations of the form $a\cdot x+b=c$ have such solutions. From the integers one can construct a more sophisticated system, the field Q of rational numbers. Formally, Q is the collection of equivalence classes of the set of all symbols m/n, where $m,n\in I^+$ and $n\neq 0$, where two symbols, m/n and m'/n', are equivalent if $mn'=m'n$. Intuitively, this means that Q is the set of all such symbols except that m/n and m'/n' represent the same rational number if $mn'=m'n$. Thus $\frac{2}{3}$, $4/6$, and $-10/-15$ all represent the same rational number. Q inherits an addition and multiplication from I; furthermore, the ordering on I induces an ordering on Q and it turns out that Q is an ordered field.

At this point let us pause to point out that Q does not satisfy the Dedekind property. Let $A=\{x\in Q\mid x^2<2 \text{ or } x \text{ is negative}\}$ and let $B=$ complement of A in Q; it is easy to show that there is no rational number x satisfying $x^2=2$; hence $Q=A\cup B$. Also, if $a\in A$ and $b\in B$, then $a<b$. Thus A and B satisfy the hypothesis of the Dedekind property. But there is no c in Q such that $a\leq c\leq b$ for all $a\in A$ and $b\in B$ because such a rational number c would satisfy $c^2=2$.

We can summarize the above, roughly, by saying that there is a "hole" in Q between the sets A and B. There are many other such holes in Q; essentially the same example with 2 replaced by any positive prime yields a hole.

The real number system is obtained by adjoining to Q enough elements to plug all the holes in Q. The elements of $R - Q$, that is, the elements which are added to Q to get R, are called *irrational numbers*. In the example we describe above, the irrational number which fills the hole between A and B is denoted $\sqrt{2}$.

It is precisely the step of going from Q to R where the construction process becomes difficult. One has the twofold problem of setting up a systematic way of locating and filling holes while simultaneously extending the algebraic and order structures so that the new system has the required properties. Let it suffice to say that this last step can be carried out and we wind up with an ordered field satisfying the Dedekind property.

The construction outlined above is purely algebraic; however, there is a way to parallel it, step by step, with a geometric one. The idea is to identify the points of a line with the elements of R.

Let L be a straight line and begin by choosing any two distinct points on L to represent the integers 0 and 1. Once these two points are chosen, the points corresponding to all the other integers are determined by marking off distances on the appropriate side of 0.

Once the points corresponding to the integers are determined, a well-known straightedge-and-compass construction can be used to locate the points corresponding to $\frac{1}{2}, \frac{1}{3}, \frac{1}{4}, \ldots$; the location of any rational is then a matter of adding distances.

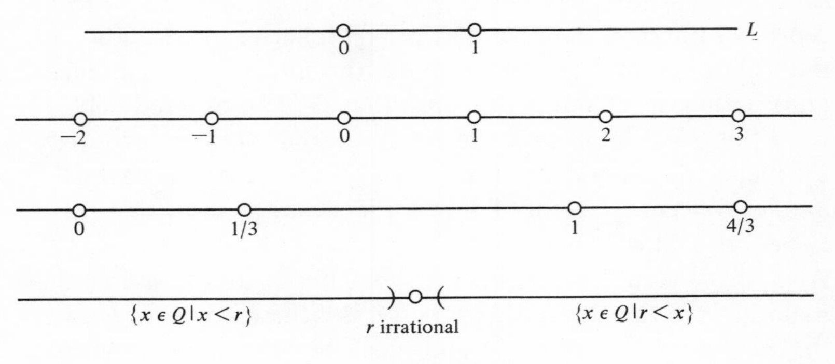

FIGURE 5.1

The last step, that of locating the irrational numbers, is already done for us. The irrational numbers are represented by precisely those points of the line which do not correspond to any rational. Some, but not all, of them can be constructed by ruler and compass, but in general the best we can say is that the point of L corresponding to the irrational number r is the point which lies between the subsets of L which represent $\{x \in Q \mid x < r\}$ and $\{x \in Q \mid r < x\}$.

Figure 5.1 and the exercises illustrate various stages in the construction process.

EXERCISES

Start off with a line and two points designated 0 and 1.

1. Show explicitly how to construct the points corresponding to 2, -4 using a compass only.

2. Using a straightedge and compass, construct $\frac{1}{2}$, $\frac{1}{5}$, $2 + (\frac{3}{5})$, and $-7 + 1\frac{1}{2}$.

3. Show how to construct the irrational numbers $\sqrt{2}$, $\sqrt{5}$, and, more generally, show how to construct any number of the form $\sqrt{a^2 + b^2}$, where $a, b \in I$.

4. Show, by induction, that, for any $n \in I^+$, $\sqrt{n}$ can be constructed.

2. LEAST UPPER BOUND PROPERTY

We begin our study of the real number system by deriving some consequences of the Dedekind property. From this point on, unless otherwise stated, "set" means "subset of R," and "point" or "element" or "number" means "element of R."

In Section 6, Chapter 3, we defined the terms upper bound, lower bound, least upper bound (l.u.b), and greatest lower bound (g.l.b.). For the usual partial ordering $<$ on R, the terms *supremum* (sup) and *infimum* (inf) are frequently used in place of "least upper bound" and "greatest lower bound." Here are two examples illustrating these concepts for subsets X of R.

2.1. $X = \{\frac{5}{3}, 2, -\frac{2}{7}, 10\}$. In this case, any number x satisfying $10 \leq x$ is an upper bound and any number y satisfying

$y \leq -\frac{2}{7}$ is a lower bound. Indeed, l.u.b. $X = 10$ and g.l.b. $X = -\frac{2}{7}$.

2.2. $X = \{1/n \mid n = 1, 2, \ldots\}$. Clearly the supremum of X is the number 1. Since all elements of X are positive, 0 is a lower bound. A little reflection leads to the guess that, in fact, inf $X = 0$. To prove this we would need to know that if $0 < a$, then a is not a lower bound for X; that is, if $0 < a$, then, for some positive integer, n, $1/n < a$. This is intuitively obvious and will be proved later in the chapter.

We are now in a position to state the first major consequence of the Dedekind property.

THEOREM 2.1 **L. U. B. Principle** *If $X \neq \varnothing$ and X is bounded above, then X has a least upper bound.*

Remark. Before proving the theorem let us note that each hypothesis is essential. If $X = \varnothing$, then X has upper bounds a plenty but no number is the least upper bound. If X is not bounded above, then it has no upper bounds and, a fortiori, no least upper bound. Finally, the theorem is certainly false if stated for a system not satisfying the Dedekind property. For example, in Q, the set $\{x \in Q \mid x^2 < 2\}$ has no least upper bound.

PROOF. Let X be nonempty and bounded above. We define two sets A and B which satisfy the hypothesis of the Dedekind property; the point c whose existence is guaranteed by the conclusion of the Dedekind property will turn out to be l.u.b. X.

Let A denote the set of points $a \in R$ such that, for some $x \in X$, $a \leq x$. Note that A contains X and is therefore nonempty.

Let $B = R - A$; a real number b is in B if and only if $x < b$ for every $x \in X$. Let b_0 be an upper bound for X; then $b_0 + 1$ is in B so B is not empty.

We leave to the reader the task of showing that if $a \in A$ and $b \in B$, then $a < b$. (It is easier to show that we cannot have $b \leq a$.)

By the Dedekind property there is a point $c \in R$ such that $a \leq c$ and $c \leq b$ for every $a \in A$ and $b \in B$. Let $x \in X$ and suppose that $c < x$, as illustrated in Figure 5.2. Observe that the number $(c + x)/2$ is in A because $(c + x)/2 < x$. But also $c < (c + x)/2$,

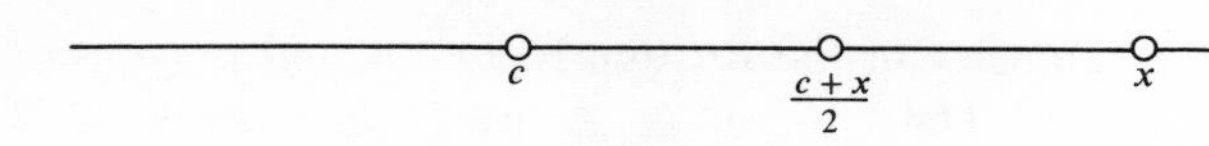

FIGURE 5.2

so this contradicts one of the requirements on c. Thus, for every $x \in X$, $x \leq c$, which says that c is an upper bound of X. Finally, suppose c' is an upper bound of X and $c' < c$. (Draw a picture.) Then it is easily seen that the number $(c' + c)/2$ is in B but is less than c, which contradicts the other condition on c. So, if c' is any upper bound of X, then $c \leq c'$. This shows that c is the least upper bound of X and the proof is complete. ▮

From the L. U. B. principle we now derive three important consequences.

THEOREM 2.2 **Archimedean Ordering Property.** *If a and b are real numbers and $a > 0$, then there is a positive integer n such that $na > b$.*

PROOF. We suppose the theorem false and derive a contradiction. Let $X = \{na \mid n \in I^+\}$; if the theorem is false, then b is an upper bound for X and, by the L. U. B. principle, X has a least upper bound, x_0. Since $a > 0$, we have $x_0 - a < x_0$, hence $x_0 - a$ is not an upper bound of X and there is some multiple of a, say $n_0 a$, such that $x_0 - a < n_0 a$. But then $x_0 < n_0 a + a = (n_0 + 1)a \in X$, which contradicts the fact that x_0 is an upper bound of X. ▮

The following result will be used time and again in studying sequential convergence (defined later) in R.

COROLLARY 2.3 *If b is a positive real number, then there is some integer n such that $1/n < b$.*

We leave the proof as an exercise.

The final result of this section could be called the rational sandwich theorem. It says that between any two different real numbers there is a rational number. This result is not too surprising if one keeps in mind the construction of R from D but from a cardinality standpoint it is interesting because the reals are uncountable and the rationals are countable.

THEOREM 2.4 **Density of the Rationals** *Let a and b be real numbers with $a<b$. Then there is a rational p such that $a<q<b$.*

PROOF. Let N be the largest integer such that $N \leq a$. If $N+1<b$, then we are done because $N+1$ is a rational and $a<N+1<b$. So we are left with the case that $N \leq a < b \leq N+1$.

Choose a positive integer n such that $1/n < b-a$ and divide the segment from N to $N+1$ into subsegments as follows. For each i, $1 \leq i \leq n$, let

$$A^i = \left\{x \mid N + \frac{i-1}{n} \leq x < N + \frac{i}{n}\right\}$$

Let A^i be that segment which contains a (Figure 5.3). Notice that, since A^i has length $1/n$ and since $1/n < b-a$, we must have $N+(i/n) < b$. Thus the rational number $q = N + i/n$ satisfies $a<q<b$. ∎

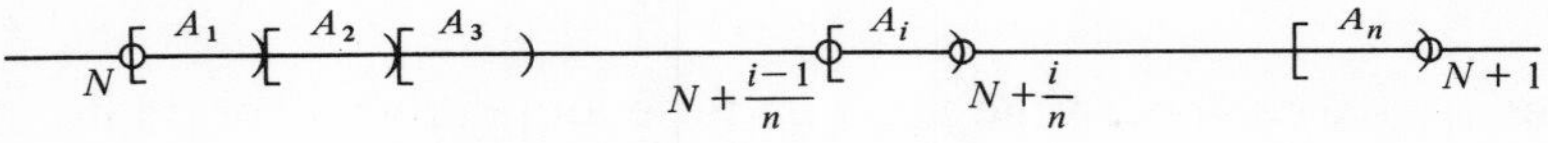

FIGURE 5.3

EXERCISES

We urge the reader to make liberal use of pictures first as an aid in solving problems and, second, to help others understand his solution.

1. Find the l.u.b. and g.l.b. of the following sets if they exist. You should convince yourself that you can prove your answer is right.

(a) $\{x \mid 0 < x \leq 1\}$.
(b) $\{x \mid 0 < x \leq 1$ and x is rational$\}$.
(c) $\{x \mid x^2 < 2\}$.

(d) The union of the sequence of sets $\{A_n\}$, where $A_n = \{x \mid 1 + (1/n) < x < n\}$.

Some of the following have already been given in Chapter 3.

2. As we remarked in Chapter 3, there is a duality between statements involving upper bounds, l.u.b.'s, and so on, and statements involving lower bounds, g.l.b.'s, and so on. State and sketch a proof of the dual of the L.U.B. Principle.

3. Prove or disprove each of the following statements.

(a) l.u.b. $(A \cup B) =$ l.u.b. {l.u.b. A, l.u.b. B}.
(b) Same statement for g.l.b.
(c) $-$l.u.b.A $=$ g.l.b. $(-A)$, where $-A = \{-x \mid x \in A\}$.

4. Prove that the number a is an upper bound for the set X if and only if for every number c such that $a < c$ we have $x < c$ for every x in X.

5. Let A be a nonempty set and suppose the set U of upper bounds of A is nonempty. Prove that l.u.b.$A =$ g.l.b. U.

6. Suppose A, B are nonempty sets bounded above and suppose, further, that given any point a of A there is a point b of B such that $a \leq b$. Show that l.u.b. $B \geq$ l.u.b. A. How can you modify this condition to get the conclusion that l.u.b. B. $=$ l.u.b. A?

7. (a) Prove the corollary to the Archimedean ordering property.
(b) Prove that if $b > 0$, then, for some positive integer n, $\sqrt{2}/n < b$.
(c) Prove that if a and b are real numbers with $a < b$, then there is a number q of the form $N + (i\sqrt{2}/n)$, where N, i, and n are integers, such that $a < q < b$.

3. SEQUENTIAL CONVERGENCE IN R

As the title of this section suggests, we are going to study sequences of real numbers in this section. This, in turn, will provide the machinery we need to study R itself. For the moment the order and distance structures on R will be of primary importance.

Before getting to the main topic of this section we introduce some new definitions and notation. An *open interval* is a set of one of the forms

$$(a, b) = \{x \mid a < x < b\},$$
$$(-\infty, b) = \{x \mid x < b\},$$
$$(a, \infty) = \{x \mid a < x\}.$$

Sometimes we also consider R itself as an open interval. The points a and b are called the *end points* of (a, b). The symbol "∞" is read "infinity"; it is not a real number.

Let $\{x_n\}$ be a sequence in R and A a subset of R. We say that $\{x_n\}$ is *ultimately in A* provided there is a positive integer N such that, if $n \geq N$, then $x_n \in A$. Thus the following statements are equivalent:

1. $\{x_n\}$ is ultimately in A.
2. All but a finite number of terms of $\{x_n\}$ are in A.
3. At most finitely many terms of $\{x_n\}$ are not in A.

Here is the main definition of this section. Let $\{x_n\}$ be a sequence in R and x a point of R. We say that $\{x_n\}$ *converges* to x provided that if A is any open interval containing x, then $\{x_n\}$ is ultimately in A.

We emphasize that in order for $\{x_n\}$ to converge to x, $\{x_n\}$ must ultimately be in *every* open interval containing x, not just some of the open intervals which contain x. At the same time it is obvious that we need only consider intervals of the form (a, b), since every open interval containing x contains an interval of this form which contains x.

Here are some examples.

Example 1. *The sequence* $\{1/n\}$ *converges to* 0. Let A be any open interval containing 0 and choose a and b such that $0 \in (a, b) \subseteq A$. Now we have $0 < b$; hence by the corollary to the Archimedean ordering property there is some $N \in I^+$ such that $1/N < b$. Then for every $n > N$ we have $0 < 1/n \leq 1/N < b$, and therefore $\{1/n\}$ is ultimately in (a, b) and, a fortiori, ultimately in A.

Suppose in Example 1 we consider two specific intervals containing 0, say $A = (-\frac{1}{100}, \frac{1}{100})$ and $B = (-\frac{1}{1000}, \frac{1}{1000})$. Now the integer $N = 101$ "works" for A in the sense that if $n \geq 101$, then $1/n \in A$. On the other hand, the smallest integer which "works" for B is $N = 1001$. This reflects what happens in almost all nontrivial situations; roughly stated, the smaller the interval, the larger the N.

Example 2. *The sequence* $\{1/n\}$ *does not converge to* $\frac{1}{1000}$. What we must show here is that there is at least one open interval containing $\frac{1}{1000}$ such that $\{1/n\}$ is not ultimately in this interval. This is a simple matter of making the right choice; we take $A = (\frac{1}{1001}, \frac{1}{999})$. Then exactly one term of the sequence is in A; hence the sequence is not ultimately in A and does not converge to $\frac{1}{1000}$.

In Example 2 we could just have easily have proved that

$\{1/n\}$ does not converge to $\frac{1}{75}$, $\frac{1}{100}$, or, indeed, to any number different from 0. This illustrates a general phenomenon, which we state as follows.

THEOREM 3.1 *If $\{x_n\}$ converges to x and $y \neq x$, then $\{x_n\}$ does not converge to y.*

PROOF. Since $y \neq x$ we have either $x < y$ or $y < x$. Supposing that $x < y$, let $A = ((x + y)/2, y + 1)$. Then A is an open interval containing y. Let $B = (x - 1, (x + y)/2)$; then, at most finitely many terms of $\{x_n\}$ are not in B, hence at most finitely many are in A, and $\{x_n\}$ is not ultimately in A. Similarly, if $y < x$, then there is an open interval A containing y such that $\{x_n\}$ is not ultimately in A; so in either case $\{x_n\}$ does not converge to y. ▮

If $\{x_n\}$ converges to x we say that x is the *limit* of the sequence $\{x_n\}$ and we write $x_n \to x$ or $\lim x_n = x$. If $\{x_n\}$ converges to some (unnamed) point we say that $\{x_n\}$ is *convergent*. The preceding result says that a convergent sequence has exactly one limit.

The next result is intuitively obvious but important.

THEOREM 3.2 *If $\{x_n\}$ converges to x, then so does every subsequence of $\{x_n\}$.*

PROOF. Let $\{y_n\}$ be a subsequence of $\{x_n\}$, say $y_n = x_{f(n)}$, where f is a selection function. Let (a, b) be an open interval containing x and choose $N \in I^+$ such that if $n \geq N$, then $x_n \in (a, b)$. We assert that the integer N also "works" for the subsequence $\{y_n\}$. More precisely, if $n \geq N$, then we have $y_n = x_{f(n)}$ and $f(n) \geq f(N) \geq N$ (the second inequality follows from Exercise 2, Section 5, Chapter 2); thus $y_n \in (a, b)$. ▮

We are now going to give some conditions on a sequence which imply that it converges. Such conditions are called convergence criteria. As usual, we have some preliminaries before stating the result.

We say that a sequence $\{x_n\}$ is *bounded* provided its range is a bounded set in R, that is, provided there exist points a and b such that $a \leq x_n \leq b$ for every $n \in I^+$. If a sequence is bounded, then the l.u.b. (g.l.b.) of the sequence is the l.u.b. (g.l.b.) of its range.

We now examine the relationship between boundedness and convergence. First, it is clear that:

1. *A Convergent Sequence is Bounded.* Suppose, say, that $x_n \to x$. Choose an integer N such that, if $n \geq N$,

$$x_n \in (x - 1, x + 1).$$

The finite set $\{x_1, x_2, \ldots, x_n\}$ has a g.l.b., say a, and a l.u.b., say b. Let a' be the smaller of a and $x - 1$ and let b' be the larger of $x + 1$ and b. Then, for every $n \in I^+$ we have $a' \leq x_n \leq b'$.

It is equally clear that:

2. *A Bounded Sequence Need Not Converge.* An example is furnished by the sequence $\{(-1)^n\}$ (see Exerise 5). One reason that this sequence fails to converge is that it hops back and forth from -1 to 1, and hence it is not ultimately in any interval whose length is less than (say) 1. We add to boundedness a condition which outlaws such "hopping around" to obtain our convergence criterion.

We say that $\{x_n\}$ is *nondecreasing* provided that if $n < m$, then $x_n \leq x_m$, and *nonincreasing* provided that if $n < m$, then $x_n \geq x_m$. The reader should look at some examples. A sequence is called *monotone* if it is nonincreasing or nondecreasing (or both).

THEOREM 3.3 *A sequence which is bounded and monotone converges. More precisely:*

(a) *A bounded nondecreasing sequence converges to its l.u.b.*

(b) *A bounded nonincreasing sequence converges to its g.l.b.*

PROOF. We prove (a); the proof of (b) is obtained by dualizing this argument. Suppose that $\{x_n\}$ is bounded, nondecreasing, and let $x = $ l.u.b. $\{x_n\}$. (Why does $\{x_n\}$ have a l.u.b.?) Let (a, b) be an open interval containing x; then $a < x$ and therefore there is some $N \in I^+$ such that $a < x_N$ (why?). But then, for every integer $n > N$ we have both $a < x_n$ and $x_n < b$ (why?); hence $\{x_n\}$ is ultimately in (a, b). ∎

We could now begin to test some specific sequences for convergence; however, at our present stage, a separate argument would have to be given for each sequence and the process would be tedious and, perhaps, a waste of time. Therefore, we shall postpone looking at any complicated examples until after Section 4, by which time plenty of machinery will be set up.

EXERCISES

1. Prove that the intersection of two open intervals is an open interval (if it is nonempty). Is the same true of an arbitrary collection of open intervals?

2. Let $\mathscr{C}$ be a collection of open intervals such that $\cap\mathscr{C} \neq \varnothing$. Show that $\cup\mathscr{C}$ is an open interval.

3. Let $S_1, \ldots, S_4$ denote the following sequences: $\{1/n\}$, $\{(-1)^n\}$, $\{1 - \frac{1}{2}n\}$, $\{-1 - \frac{1}{2}n\}$; and let A_1, A_2, A_3, and A_4 be the intervals $(-\frac{4}{3}, 0)$, $(-\frac{7}{6}, 2)$, $(\frac{1}{2}, 2)$, and $(-2, -\frac{1}{2})$. For each i and j $(1 \leq i, j \leq 4)$ answer the question: Is S_i ultimately in A_j?

4. We might say that $\{x_n\}$ is *ultimately out of A* in case $\{x_n\}$ is ultimately in $R - A$. Is this the same as saying that $\{x_n\}$ is not ultimately n A?

5. Verify each of the following statements.
(a) $-1/n \to 0$.
(b) $\dfrac{(-1)^n}{n} \to 0$.
(c) $\{(-1)^n\}$ is not convergent.
(d) A constant sequence (all terms equal) converges.
(e) $\{n\}$ is not convergent.
(f) $\left\{\dfrac{3^2}{5^{n+1}}\right\}$ converges.
(g) $\left\{\dfrac{n^2}{2^n}\right\}$ converges.
(*Hint*: Most of these can be disposed of by appealing to one or more of the results of this section. Others require a little arguing.)

6. Prove this theorem: Suppose $\{x_n\}$ converges to x and $\{y_n\}$ is a sequence such that for some integer $N \in I^+$, if $n \geq N$, then $y_n = x_n$ (that is, $\{x_n\}$ and $\{y_n\}$ are ultimately the same). Then $\{y_n\}$ converges to x.

4. ALTERNATIVE DEFINITION OF CONVERGENCE

We begin this section with another version of the definition of convergence. This will be stated in terms of the distance between points of R, so we begin by reviewing some of the relevant definitions and facts.

If a and b are in R, then the distance from a to b is the absolute value of the difference of a and b; that is, the distance from a to b is $|a - b|$. For any real numbers a, b, and c, the following are true:

4.1. $|a - b| \geq 0$ and $= 0 \Leftrightarrow a = b$.
4.2. $|a - b| = |b - a|$.
4.3. $|a - b| \leq |a - c| + |c - b|$.

The third statement is known as the triangle inequality and, in one form or another, will be used often in what follows.

We now state, as a theorem, the alternative definition of convergence.

THEOREM 4.1 *Let $\{x_n\}$ be a sequence and x a point. The following are equivalent:*

(a) $x_n \to x$.
(b) *Corresponding to every positive real number ε there is an integer N such that if $n \geq N$, then $|x_n - x| < \varepsilon$.*

PROOF. Suppose $x_n \to x$ and let $\varepsilon > 0$ be given. The sequence $\{x_n\}$ is ultimately in the open interval $(x - \varepsilon, x + \varepsilon)$. Thus there is an integer N such that if $n \geq N$, then $x - \varepsilon < x_n < x + \varepsilon$. This last statement can be rewritten in the form $|x_n - x| < \varepsilon$; hence (b) holds.

Conversely, suppose (b) holds and let (a, b) be any open interval containing x. Let ε be the smaller of the numbers $x - a$, $b - x$ (Figure 5.4). Then ε is positive and, by (b), there is

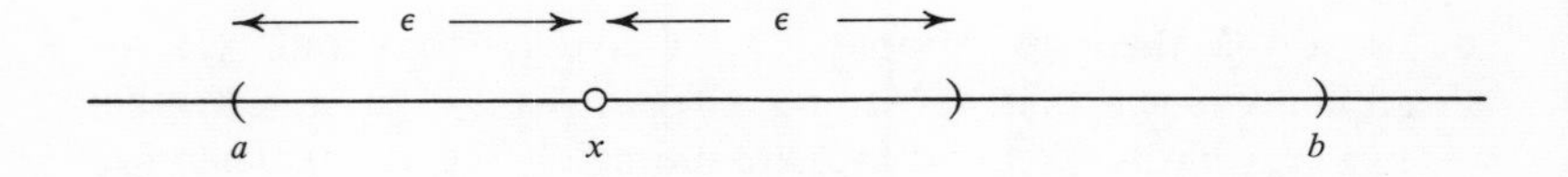

FIGURE 5.4

$N \in I^+$ such that if $n \geq N$, then $|x_n - x| < \varepsilon$. But this implies that if $n \geq N$, then $a < x_n < b$; that is, $\{x_n\}$ is ultimately in (a, b). ∎

Remarks. As noted in the above proof, statement (b) is merely a

long-winded way of saying that for every positive ε, $\{x_n\}$ is ultimately in the interval $(x - \varepsilon, x + \varepsilon)$.

One effect of this theorem is to allow us to use intervals centered at x rather than arbitrary intervals containing x in the definition of convergence. A second, more important, effect is to permit full use of the properties of the absolute-value function in investigating convergence. We give some examples to illustrate this, and we also suggest that the reader try to prove some of the following results using the open interval definition of convergence rather than the $\varepsilon - N$ definition.

Example 1. Let $x_n = \sin n/n$; we assert that $x_n \to 0$. To see this, suppose $\varepsilon > 0$ is given. Choose $N \in I^+$ such that if $n \geq N$, then $1/n < \varepsilon$. Then if $n \geq N$ we have

$$\left| \frac{\sin n}{n} - 0 \right| = \frac{|\sin n|}{n} \leq \frac{1}{n} < \varepsilon \quad \text{(why?)}$$

as desired.

Example 2. If $\{x_n\}$ is a sequence converging to x, then $\{-x_n\}$ converges to $-x$. This follows from the fact that

$$|-x_n - (-x)| = |x_n - x|.$$

Example 3. The sequence $\{(1 - 3^n)/3^n\}$ converges to -1. This follows from the fact that $|((1 - 3^n)/3^n) + 1| = 1/3^n$ and the fact that $1/3^n$ converges to 0.

The next result tells how the algebraic operations in R interact with convergence.

THEOREM 4.2 *Suppose $x_n \to x$ and $y_n \to y$. Then*

(a) $x_n + y_n \to x + y$.
(b) $x_n \cdot y_n \to x \cdot y$.
(c) $Kx_n \to Kx$ (K a constant).
(d) If $y \neq 0$ and if, for every n, $y_n \neq 0$, then $x_n/y_n \to x/y$.

Remark. The first assertion is shorthand for the more precise statement: The sequence whose nth term is $x_n + y_n$ converges to $x + y$. Similar interpretations are to be given to the other parts of the theorem.

We shall prove parts (b) and (d). Part (a) is a moderately easy exercise and part (c) follows easily from (b).

PROOF OF (b). Suppose $x_n \to x$ and $y_n \to y$. Choose numbers P and Q as follows. First, the sequence $\{x_n\}$ is bounded and we take $P > 0$ large enough so that $-P < x_n < P$ for all n; thus, for all n, $|x_n| < P$. Second, choose Q so that $|y| < Q$ ($Q = |y| + 1$ will do).

Now to prove that $x_n \cdot y_n \to x \cdot y$, suppose $\varepsilon > 0$ is given. Choose $N_1 \in I^+$ so that

$$(1) \quad |x_n - x| < \frac{\varepsilon}{2Q} \qquad \text{for all } n \geq N_1;$$

and choose $N_2 \in I^+$ so that

$$(2) \quad |y_n - y| < \varepsilon/2P \qquad \text{for all } n \geq N_2.$$

Let N be the larger of N_1 and N_2; if $n \geq N$, then (1) and (2) hold simultaneously and we have

$$\begin{aligned} |x_n y_n - xy| &= |(x_n y_n - x_n y) + (x_n y - xy)| \\ &\leq |x_n y_n - x_n y| + |x_n y - xy| \\ &= |x_n|\,|y_n - y| + |y|\,|x_n - x| \\ &< P \cdot \frac{\varepsilon}{2P} + Q \cdot \frac{\varepsilon}{2Q} = \frac{\varepsilon}{2} + \frac{\varepsilon}{2} = \varepsilon. \end{aligned}$$

(The reader should furnish a reason for each step in this chain.) Thus, if $n \geq N$, then $|x_n y_n - xy| < \varepsilon$, as required.

The best way to understand this proof is to start at the end of the chain of equalities and inequalities, and, working backward, see why we used $\varepsilon/2D$ in (1) and $\varepsilon/2P$ in (2) and why we needed P and Q in the first place.

PROOF OF (d). We shall prove that if $y \neq 0$ and $y_n \neq 0$ for every n, then $1/y_n \to 1/y$. This together with part (b) proves part (d).

Suppose, for the moment, that we can find a positive number P such that $P < |y|$ and $P < |y_n|$ for every n. Notice, then, that for every $n \in I^+$ we have $|yy_n| > P^2$, and hence $1/|yy_n| < 1/P^2$.

Given $\varepsilon > 0$, choose $N \in I^+$ such that if $n \geq N$, then $|y_n - y| < \varepsilon P^2$. Then if $n \geq N$ we also have

$$|1/y_n - 1/y| < \left|\frac{y_n - y}{yy_n}\right| < \frac{1}{|yy_n|} \cdot |y_n - y| < \frac{1}{P^2}|y_n - y| < \frac{1}{P^2} \cdot \varepsilon P^2 = \varepsilon,$$

as desired. Thus it remains to prove that such a number P exists. To see this choose $N \in I^+$ so that if $n \geq N$, then y_n lies in the open interval of length $|y|/2$ centered at y. Since each of the numbers $|y_1|, |y_2|, \ldots, |y_{N-1}|, |y|/2$, is positive, there is a number P which is less than each of them. This number has the desired properties. ▮

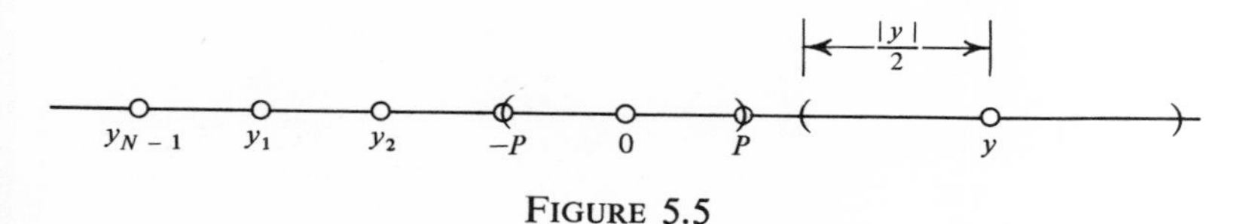

FIGURE 5.5

Figure 5.5 illustrates the situation when $0 < y$. Here all terms after the $(N-1)$st lie in the interval centered at y and P is chosen smaller than $|y|/2$ and also small enough so that none of $x_1, \ldots, x_{N-1}$ lies in $(-P, P)$.

We illustrate the use of Theorem 4.2 with two examples.

Example 1. Let r be any number such that $-1 < r < 1$. Then the sequence $\{r^n\}$ converges to 0. Let us first prove this for the case that $0 \leq r$. Since $r < 1$ we have $r^n < r^m$ if $n > m$, so $\{r^n\}$ is nonincreasing and (clearly) bounded. Thus $\{r^n\}$ converges to some number, say x.

Now consider the sequence $\{r^{2n}\}$; by part b) of the theorem this converges to x^2. But $\{r^{2n}\}$ is a subsequence of $\{r^n\}$ (what is the selection function?); hence we also know that $\{r^{2n}\}$ converges to x. So we have $x^2 = x$ and either $x = 1$ or $x = 0$. The first possibility is ruled out by the fact that, for every $n \varepsilon I^+$, $r^n \leq r < 1$.

In the case that $-1 < r \leq 0$ we know by what we just proved that $|r|^n \to 0$. Since $|r^n - 0| = |r^n| = |r|^n$, it follows that $r^n \to 0$.

Example 2. The sequence $x_n = [(3/5)^n - 7]/3 - (5/7)^n$ converges to $-7/3$. To see this, write $x_n = y_n/z_n$, where $y_n = (3/5)^n - 7$ and $z_n = 3 - (5/7)^n$. Now write $y_n = u_n + v_n$, where $u_n = (3/5)^n$ and $v_n = -7$. By the previous example, $u_n \to 0$. Also, v_n, being a constant sequence with each term equal to -7, converges to -7. Hence $y_n \to 0 - 7 = -7$. Similarly, $z_n \to 3 - 0 = 3$. Thus $x_n = y_n/z_n \to -\frac{7}{3}$.

EXERCISES

1. Prove parts (a) and (c) of Theorem 4.2.

2. Investigate each of the following sequences for convergence (the nth term of each of the sequences is given):

(a) $n/(n+1)$
(b) $(n^2-2)/(n^2+4)$
(c) $(a-bn)/(c-dn)$
(d) $(n^2-n^3)/n^3$
(e) $n^{1/n}$
(f) a^n (a a fixed real number)
(g) $(\sqrt{n+1}-\sqrt{n})/n$
(h) $(2n^2-3n+6)/(n^2-5n+6)$

3. If the sequence $\{x_n\}$ converges to x, prove that each of the following sequences $\{y_n\}$ and $\{z_n\}$ also converges to x:
$y_n = (x_1 + x_2 + \cdots + x_n)/n$ and $z_n = (x_1 \cdot x_2 \cdot \ldots \cdot x_n)^{1/n}$.

4. We say a sequence is *divergent* if it is not convergent. Prove or disprove each of the following statements.

(a) If $\{x_n\}$ is divergent it is not bounded.
(b) The converse of (a).
(c) If $\{x_n\}$ is convergent and $\{x_n y_n\}$ is divergent, then $\{y_n\}$ is divergent.
(d) If $\{x_n\}$ and $\{y_n\}$ are divergent, so is $\{x_n + y_n\}$.
(e) Any sequence is a subsequence of a divergent sequence.

5. We say $\{x_n\}$ diverges to $+\infty(-\infty)$, and write $x_n \to +\infty(-\infty)$, provided that for any number a, $\{x_n\}$ is ultimately in (a, ∞) $[(-\infty, a)]$.

(a) Show that if $x_n \to +\infty$, then $\{x_n\}$ is divergent.
(b) Show that if $x_n \to +\infty$ and $x_n \neq 0$ for all n, then $1/x_n \to 0$.
(c) Show that if x_n is nonincreasing, converges to 0, and $x_n \neq 0$ for all n, then $1/x_n \to +\infty$.
(d) Show that $x_n \to +\infty \Leftrightarrow -x_n \to -\infty$.
(e) Discuss the assertion: If $x_n \to +\infty$ and $y_n \to -\infty$, then $x_n + y_n \to 0$.

5. LIMIT POINTS OF SETS

This section constitutes somewhat of a digression. We introduce a new concept, that of a limit point of a set, and prove three theorems. What we do here will provide the machinery needed to derive the most important properties of the real number system.

Let x be a point and X a subset of R. We say x is a *limit point* of X provided that every open interval which contains x contains a point of X different from x.

Example 1. 0 is a limit point of the set $\{1/n \mid n \in I^+\}$.

Example 2. 0 is not a limit point of I (the set of integers). In fact, I has no limit points.

Example 3. Every point of R is a limit point of Q (the set of rationals) and of $R-Q$ (the set of irrationals.). These follow from the density of the rationals and irrationals.

Example 4. Every point of the open interval (a, b) is a limit point of (a, b), a and b are also limit points of (a, b), but no other point is.

It is obvious from the use of term "limit point" and from its definition that this concept is related to that of the limit of a sequence. This relationship is stated as our first result.

THEOREM 5.1 *The point x is a limit point of the set X if and only if there is a sequence in $X - \{x\}$ converging to x.*

PROOF. Let x be a limit point of X. For each integer $n \in I^+$, let x_n be a point of X different from x which lies in the interval $(x - 1/n, x + 1/n)$. Then $\{x_n\}$ is a sequence in $X - \{x\}$ converging to x. ∎

The proof of the reverse implication is left as an exercise.

Our next result gives conditions under which a set has a limit point. Clearly no finite set has any limit points, so the only candidates for sets with limit points are infinite sets. Some other condition is also required since I^+ has no limit points.

THEOREM 5.2 *A bounded infinite set has at least one limit point.*

Before giving the proof let us introduce some useful terminology. If a and b are points of R, then $[a, b]$ denotes the set

$\{x \mid a \leq x \leq b\}$; such a set is called a *closed interval* (with end points a and b).

Now let X be a bounded infinite set. Since X is bounded there exist points, a, b with $a < b$ such that $X \subseteq [a, b]$.

We now define a sequence of closed intervals, each containing the next, as follows.

Let I_1 be one of the intervals $[a, (a+b)/2]$, $[(a+b)/2, b]$ which contains infinitely many points of X. For convenience, relabel the end points of I_1 by writing $I_1 = [l_1, r_1]$.

Let I_2 be one of the intervals $[l_1, (l_1 + r_1)/2]$, $[(l_1 + r_1)/2, r_1]$ which contains infinitely many points of X and relabel end points so that: $I_2 = [l_2, r_2]$.

Continue in this way. In general, $I_n = [l_n, r_n]$ contains infinitely points of X. I_{n+1} is chosen to be one of the intervals $[l_n, (l_n + r_n)/2]$, $[(l_n + r_n)/2, r_n]$ which contains infinitely many points of X and is written $I_{n+1} = [l_{n+1}, r_{n+1}]$.

Associated with the sequence $\{I_n\}$ we have the sequence $\{l_n\}$ of left-hand end points and the sequence $\{r_n\}$ of right-hand end points. Notice that $\{l_n\}$ is nondecreasing and bounded, hence converges to its l.u.b., which we denote x.

We now assert that x is a limit point of X. Let A be any open interval containing x and let ε be a positive number small enough so that $(x - \varepsilon, x + \varepsilon) \subseteq A$. Choose a positive integer N large enough so that $x - l_N < \varepsilon$ and $b - a/2^N < \varepsilon$. By definition, I_N has length $b - a/2^N$, hence $r_N = l_N + [(b - a)/2^N]$.

It follows (see Figure 5.6) that both l_N and r_N are in

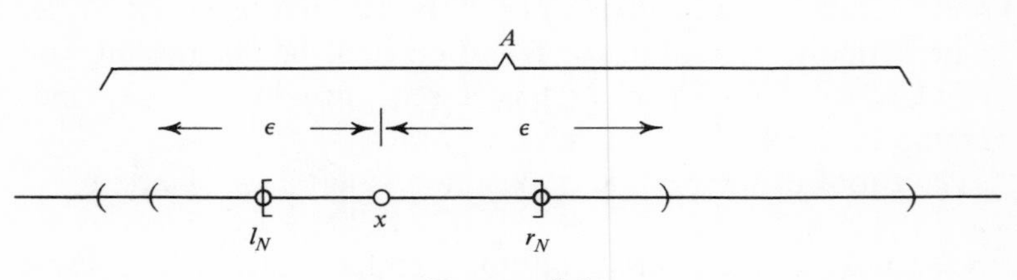

FIGURE 5.6

$(x - \varepsilon, x + \varepsilon)$, and therefore I_N is contained in A. Since I_N contains infinitely many points of X, it (and hence A) contains one different from x. Thus every open interval containing x contains a point of X different from x; that is, x is a limit point of X. ∎

We remark that in the above proof it is also possible to show that l.u.b. $l_n =$ g.l.b.r_n, using the fact that $r_n = l_n + [(b - a)/2^n]$.

We end this section by applying the preceding theorem to get a result which will be used in Chapter 6 to prove the fundamental theorem of sequential convergence.

THEOREM 5.3 *Every bounded sequence has a convergent subsequence.*

PROOF. Suppose $\{x_n\}$ is a bounded sequence. If the range of $\{x_n\}$ is finite, then, by Exercise 5, Section 5, Chapter 2, $\{x_n\}$ has a constant (and hence convergent) subsequence. If the range X of $\{x_n\}$ is infinite, then by the preceding theorem it has a limit point x. In this case we define a subsequence of $\{x_n\}$ with the following selection function.

Let $f(1)$ be a positive integer such that $|x_{f(1)} - x| < 1$. Assuming that $f(1), f(2), \ldots, f(n-1)$ have been defined, define $f(n)$ to be the first integer i greater than $f(n-1)$ such that $|x_i - x| < 1/n$. (There is such an integer, since otherwise x cannot be a limit point of X.)

From the way f is defined it is order preserving, and putting $y_n = x_{f(n)}$, we get a subsequence $\{y_n\}$ such that, for each n, $|y_n - x| < 1/n$; thus $y_n \to x$. ∎

Theorems like the one we just proved are called existence theorems. If often happens that such a theorem, while guaranteeing the existence of some object (a convergent subsequence in this case), sheds little or no light on how we may actually exhibit the object.

For example, consider the sequence $\{x_n\}$, where $x_n = \sin n$ (radian measure). This sequence is bounded by -1 and 1, so it has a convergent subsequence. In fact, this sequence has many convergent subsequences, but the theorem does not tell us explicitly how to exhibit even one of them. The reader might wish to examine a few terms hoping to find a pattern which will suggest a candidate for a convergent subsequenee. Here are approximations to the first seven terms: $x_1 \approx .84$, $x_2 \approx .91$, $x_3 \approx .14$, $x_4 \approx -.74$, $x_5 \approx -.96$, $x_6 \approx -.29$, and $x_8 \approx .63$.

It is not our intention to disparage existence theorems; they are very powerful theoretic tools. The above remarks merely illustrate the fact that it is possible to know mathematically that an object exists without being able to display it.

EXERCISES

1. Verify Examples 1 through 4. Show also that no finite set has a limit point.

2. The moral of this problem is that there is a distinction between being the limit of a sequence and being a limit point of its range.

(a) Give an example of a sequence $\{x_n\}$ converging to a point x such that x is not a limit point of the range of $\{x_n\}$.

(b) Give an example of a sequence $\{x_n\}$ in R such that every point of R is a limit point of the range of $\{x_n\}$.

Prove that the sequence of example (b) cannot converge to any point of R by proving

(c) If $x_n \to x$ and y is a limit point of the range of $\{x_n\}$, then $x = y$.

3. Show that if x is a limit point of the range of $\{x_n\}$ then there is a subsequence of $\{x_n\}$ which converges to x. (*Remark*: Theorem 5.1 does not imply this result although its proof can be modified to do do so.) Illustrate your proof using the sequence $x_n = (-1)^n$.

4. (a) Give several examples of existence theorems.

(b) A uniqueness theorem is one which asserts that if there is at least one object with certain properties, then there is exactly one. Give some examples of uniqueness theorems.

6. CAUCHY SEQUENCES

As we found in Section 3, a bounded sequence which is monotone converges. Certainly the converse is false; that is, a convergent sequence need not be monotone. Thus, for bounded sequences, monotoniety is a sufficient but not a necessary condition for convergence. In this section we find a condition which is both necessary and sufficient for convergence.

We say that a sequence $\{x_n\}$ is *Cauchy* provided that it satisfies the following condition:
(*) Given a positive number ε, there is an integer $N \in I^+$ such that

$$\text{if } n \geq N \text{ and } m \geq N, \text{ then } |x_n - x_m| < \varepsilon.$$

We might paraphrase condition (*) by saying that, given any

$\varepsilon > 0$, the distance between terms of $\{x_n\}$ is ultimately less than ε.

We omit giving examples of Cauchy and non-Cauchy sequences since the reader can supply his own once he believes the following result.

THEOREM 6.1 **Cauchy Criterion for Convergence** *A sequence converges if and only if it is Cauchy.*

PROOF. Let $\{x_n\}$ be Cauchy; we show first that $\{x_n\}$ has a convergent subsequence, say $y_n \to x$, and then that $\{x_n\}$ itself converges to x.

Choose an integer $N \in I^+$ such that for $n, m \geq N$ we have $|x_n - x_m| < 1$ (we have taken $\varepsilon = 1$). Thus, in particular, for all $n \geq N$ we have $x_n \in (x_N - 1, x_N + 1)$. The finite set

$$\{x_1, x_2, \ldots, x_{N-1}\}$$

has a g.l.b., a, and a l.u.b., b. Let a' be the smaller of $x_N - 1$ and a and let b' be the larger of $x_N + 1$ and b; then, for every $n \in I^+$, $a' \leq x_n \leq b'$. Thus the sequence $\{x_n\}$ is bounded and has a convergent subsequence, say $y_n \to x$. Let f be the corresponding selection function; thus $f\colon I^+ \to I^+$ is order preserving and, for every n, $y_n = x_{f(n)}$.

We now show that $x_n \to x$. To see this, suppose $\varepsilon > 0$ is given. Choose N_1 such that if $n,m \geq N_1$, then $|x_n - x_m| < \varepsilon/2$ and choose $N \geq N_1$ such that $|y_N - x| < \varepsilon/2$. Notice that $f(N) \geq N$ (Exercise 2, Section 5, Chapter 2); hence $f(N) \geq N_1$. Now, at last, suppose that $n \geq N$; then $|x_n - x| \leq |x_n - x_{f(N)}| + |x_{f(N)} - x|$. The first term is less than $\varepsilon/2$ because $n \geq N_1$ and $f(N) \geq N_1$. The second term, which can be rewritten $|y_N - x|$, is less than $\varepsilon/2$ by choice of N. Thus, if $n \geq N$, then $|x_n - x| < \varepsilon$ and we have shown that $\{x_n\}$ converges to x.

This proves one half of the theorem. The converse, which says that a convergent sequence is Cauchy, is left to the reader as an exercise. The Cauchy criterion will be used in this book to help study infinite series. The reader will see it used again in more advanced courses. It is so important that very often in the course of constructing exotic spaces based on the real number system mathematicians require that the new space at least satisfy the Cauchy criterion or some vestigial form of it.

EXERCISES

1. Prove that if $\{x_n\}$ is convergent, then it is Cauchy.

2. Call a subset X of R *complete* provided that every Cauchy sequence in X converges to a point of X. Which of the following sets are complete and why?

   (a) R (b) $[0, 1] = \{x \mid 0 \leq x \leq 1\}$
   (c) $(0, 1)$ (d) $\{1/n \mid n \in I^+\}$
   (e) $\{x \mid x \leq 0\} \cup \{x \mid 1 \leq x\}$ (f) $\{1 - 1/2n \mid n \in I^+\} \cup \{1\}$

3. Prove or disprove each of the following.

   (a) If $\mathscr{A}$ is a finite collection of complete subsets of R, then $\cap \mathscr{A}$ is complete.
   (b) If $\mathscr{A}$ is a finite collection of complete subsets of R, then $\cup \mathscr{A}$ is complete.
   (c) If $\mathscr{A}$ is any collection of complete subsets of R, then $\cap \mathscr{A}$ is complete.
   (d) If $\mathscr{A}$ is any collection of complete subsets of R, then $\cup \mathscr{A}$ is complete.

4. Let X be a complete subset of R and let x_0 be any element of R. Must the set $x_0 + X = \{x_0 + x \mid x \in X\}$ be complete? Must the set $x_0 \cdot X = \{x_0 \cdot x \mid x \in X\}$ be complete?

5. Show that if X is any subset of R and x is a limit point of X then $X - \{x\}$ is not complete.

6. Show that a subset X of R is complete if and only if X contains all its limit points.

7. CLOSED SETS, OPEN SETS, AND COMPACTNESS

In this section we expand some of the ideas developed in Section 5. We say that a subset X of R is *closed* provided that X contains all its limit points.

Thus, for example, a closed interval $[a, b]$ is a closed set. A set which has no limit points is automatically closed; so any finite set (including the null set) is closed, as is the set of all integers. The open interval (a, b) and the set Q of rational numbers are not closed.

We now state what might be called the sequential characterization of closed sets.

THEOREM 7.1 *The following are equivalent for a subset X of R:*

(a) *X is closed.*
(b) *If $\{x_n\}$ is a sequence in X and $x_n \to x$, then $x \in X$.*

The proof is left as an exercise in translating the corresponding result of Section 5.

The next result discusses intersections and unions of closed sets.

THEOREM 7.2 *Let $\mathscr{F}$ be a collection of closed sets. Then $\cap \mathscr{F}$ is closed and, if $\mathscr{F}$ is finite, $\cup \mathscr{F}$ is closed.*

PROOF. Let us denote $\cap \mathscr{F}$ by X. Suppose $\{x_n\}$ is a sequence in X converging to x. Then, for each F in $\mathscr{F}$, $\{x_n\}$ is a sequence in F and, since F is closed, x belongs to F. Thus x is in every member of $\mathscr{F}$, hence x is in $\cap \mathscr{F} = X$.

Next suppose $\mathscr{F}$ is finite and $\{x_n\}$ is a sequence in $X = \cup \mathscr{F}$ converging to x. Some member F of $\mathscr{F}$ contains infinitely many terms of $\{x_n\}$, and it is an easy matter to construct a subsequence $\{y_n\}$ of $\{x_n\}$ which is in F. Since $\{y_n\}$ must also converge to x and since F is closed, x belongs to F and hence to $\cup \mathscr{F}$. ▮

We call a set X *open* provided its complement $R - X$ is closed. Thus the complement of any finite set is open and the empty set and R itself are open. Any open interval (a, b) is open because its complement is the union of two closed sets $\{x \mid x \le a\}$, $\{x \mid b \le x\}$ and hence is closed. There are, of course, sets which are neither open nor closed; the set of rational numbers is an example.

Because of the definition of "open" in terms of "closed," there is a sort of duality between statements about open sets and statements about closed sets. Every statement about one type of set translates into a statement about the other by "taking complements." As an illustration here is the dual of the theorem we just proved.

THEOREM 7.3 *Let $\mathscr{O}$ be a collection of open sets. Then $\cup \mathscr{O}$ is open and, if $\mathscr{O}$ is finite, then $\cap \mathscr{O}$ is open.*

PROOF. Let $\mathscr{F} = \{R - O \mid O \in \mathscr{O}\}$. Then $\mathscr{F}$ is a collection of closed sets. Now $\cap \mathscr{F}$ is closed, hence $R - \cap \mathscr{F}$ is open. By the

DeMorgan formulas we have $R - \cap \mathscr{F} = \cup\{R - F \mid F \in \mathscr{F}\} = \cup\{R - (R - O) \mid O \in \mathscr{O}\} = \cup\, \mathscr{O}$. If $\mathscr{O}$ is finite, so is $\mathscr{F}$; then $\cup \mathscr{F}$ is closed and $R - \cup \mathscr{F}$ is open. Using the other DeMorgan formula, we have $R - \cup \mathscr{F} = \cap\, \mathscr{O}$. ▮

The duality between open and closed sets permits us to formulate our results for whichever class of sets is more convenient. For a while we shall concentrate on open sets.

THEOREM 7.4 *The following are equivalent for a set $X \subseteq R$:*

(a) *X is open.*
(b) *If $x \in X$, then there is an open interval containing x and contained in X.*

PROOF. Condition (b) is equivalent to the statement that no point of X is a limit point of $R - X$, which is another way of saying that $R - X$ contains all its limit points. Thus (b) holds if and only if $R - X$ is closed, that is, if and only if X is open. ▮

Before proving the final two results of the section we pause to establish a useful fact about collections of open sets.

LEMMA 7.5 *If $\mathscr{C}$ is any collection of open sets, then there is a countable subcollection $\mathscr{D}$ of $\mathscr{C}$ such that $\cup \mathscr{D} = \cup \mathscr{C}$.*

PROOF. Let $\mathscr{R}$ be the collection of all open intervals of the form (a, b), where a and b are rational numbers; $\mathscr{R}$ is countable (Exercise 2). Let $\mathscr{R}'$ denote the collection of all members A of $\mathscr{R}$ which are subsets of members of $\mathscr{C}$. Certainly $\mathscr{R}'$ is countable and we assert that $\cup \mathscr{R}' = \cup \mathscr{C}$. Suppose x is in $\cup \mathscr{C}$; pick $C \in \mathscr{C}$ such that $x \in C$. Then since C is open, there is an open interval A such that $x \in A \subseteq C$. By the density of the rationals there is a member R of $\mathscr{R}$ such that $x \in R \subseteq A$. This R is in $\mathscr{R}'$, hence $x \in \mathscr{R}'$. We have shown that $\cup \mathscr{C} \subseteq \cup \mathscr{R}'$ and the reverse containment is trivial (why?), so $\cup \mathscr{C} = \cup \mathscr{R}'$, as asserted.

Now, for each $R \in \mathscr{R}'$ pick one element C_R of $\mathscr{C}$ which contains R. Then the collection $\mathscr{D} = \{C_R \mid R \in \mathscr{R}'\}$ is countable and $\cup \mathscr{D} = \cup \mathscr{R}' = \cup \mathscr{C}$. ▮

The next result says that open sets are "generated by" open intervals. Let $\mathscr{C}$ be a collection of sets; we say the elements of $\mathscr{C}$

are *pairwise disjoint* provided that if $A, B \in \mathscr{C}$ and $A \neq B$, then $A \cap B = \varnothing$.

THEOREM 7.6 *Let X be an open set. Then there is a countable collection $\mathscr{C}$ of pairwise disjoint open intervals such that $\cup\mathscr{C} = X$.*

PROOF. For each x in X let I_x denote the union of all open intervals which contain x and lie in X. By the preceding theorem there is at least one such interval and by Exercise 2, Section 3, the union of all such intervals is an open interval.

Let $\mathscr{I}$ be the collection $\{I_x \mid x \in X\}$. If x and y are points of X and $I_x \cap I_y \neq \varnothing$, then $I_x \cup I_y$ is an open interval containing both x and y and lying in X, and hence $I_x \cup I_y$ is contained in I_x and in I_y. It follows from elementary set theory that, in this case, $I_x = I_y$.

So far, then, we have a collection $\mathscr{I}$ of pairwise disjoint open intervals such that $\cup\mathscr{I} = X$. By the lemma, there is a countable subcollection $\mathscr{C}$ of $\mathscr{I}$ such that $\cup\mathscr{C} = \cup\mathscr{I} = X$. The elements of $\mathscr{C}$ are still pairwise disjoint and we are done. ▮

We remark that, in fact, the collection $\mathscr{C}$ must contain all elements of $\mathscr{I}$; that is, $\mathscr{C} = \mathscr{I}$ and $\mathscr{I}$ was countable to start with. This is stated as Exercise 3.

Before stating the final, and most important, theorem of this section we give a definition. We say that a subset X of R is *compact* provided that if $\mathscr{C}$ is any collection of open sets whose union contains X, then there is a finite subcollection of $\mathscr{C}$ whose union covers X.

Obviously any finite set is compact. On the other hand, the set I of integers is not compact; for if we let $\mathscr{C} = \{(i - \frac{1}{2}, i + \frac{1}{2}) \mid i \in I\}$, then $\cup\mathscr{C}$ contains I but the union of any finite subcollection if $\mathscr{C}$ does not.

It turns out that for subsets of R compactness is an extremely "nice" property to have. This will be illustrated in Section 8 and in more advanced courses. For now we content ourselves with the following result.

THEOREM 7.7 *The following are equivalent for a subset X of R:*

(a) *X is compact.*

(b) *X is closed and bounded.*

(c) *Every sequence in X has a subsequence which converges to a point of X.*

The pattern of proof is (a) $\Rightarrow$ (b) $\Rightarrow$ (c) $\Rightarrow$ (a).

To prove that (a) $\Rightarrow$ (b), suppose X is compact. Consider the collection $\mathscr{C} = \{(-n, n) \mid n \in I^+\}$; certainly $\cup\mathscr{C}$ contains X. Let $\mathscr{D}$ be a finite subcollection of $\mathscr{C}$ such that $X \subseteq \cup\mathscr{D}$; then $\cup\mathscr{D}$ is an interval of the form $(-N, N)$, for some $N \in I^+$; hence $X \subseteq (-N, N)$ and X is bounded.

Now suppose x is a point of R, which is not in X; we show that x is not a limit point of X. For each integer $n \in I^+$ let O_n be the open set $R - [x - 1/n, x + 1/n]$ and let $\mathscr{C}$ be the collection of all such O_n's. Now $\cup\mathscr{C} = R - \{x\}$; hence $X \subseteq \cup\mathscr{C}$ and there is a finite subcollection $\mathscr{D}$ of $\mathscr{C}$ such that $X \supseteq \cup\mathscr{D}$. It follows that X is contained in some O_N; hence $R - O_N = [x - 1/N, x + 1/N]$ is contained in $R - X$. Thus there is an open interval which contains x and whose intersection with X is empty. This proves that x is not a limit point of X. So X contains all its limit points and is closed.

The implication (b) $\Rightarrow$ (c) is left as Exercise 4.

To prove that (c) $\Rightarrow$ (a), suppose X satisfies (c) but not (a). Let $\mathscr{C}$ be a collection of open sets with $X \subseteq \mathscr{C}$ such that for any finite subcollection $\mathscr{D}$ of $\mathscr{C}$, there is a point of X not in $\cup\mathscr{D}$. By the lemma we can choose a countable subcollection $\mathscr{C}'$ of $\mathscr{C}$ such that $\cup\mathscr{C}' = \cup\mathscr{C}$. Write $\mathscr{C}'$ as a sequence: $\mathscr{C}' = \{C_n \mid n \in I^+\}$.

Define a sequence $\{x_n\}$ in X by letting x_n be a point of $X - (C_1 \cup C_2 \cup \cdots \cup C_n)$ for each $n \in I^+$. Now, by (c), there is a subsequence $\{y_n\}$ of $\{x_n\}$ which converges to a point x of X. Let f be the corresponding selection function, so $y_n = x_{f(n)}$ for every $n \in I^+$. Now for some integer N, x belongs to C_N. Since $y_n \to x$ and C_N is open, $\{y_n\}$ is ultimately in C_N (why?) and there is an integer N' such that $N' \geq N$ and $y_{N'} \in C_N$. Now $y_{N'} = x_{f(N')}$ and $f(N') \geq N' \geq N$; but, by definition, $x_{f(N')} \notin C_1 \cup \cdots \cup C_N \cup \cdots \cup C_{f(N')}$, so we have a contradiction. Thus, if X does not satisfy (a), it does not satisfy (c); that is, (c) $\Rightarrow$ (a). ∎

EXERCISES

1. Prove the first theorem of the section.
2. Show that the collection of all open intervals with rational end points is countable.

3. Show that if $\mathscr{C}$ is a collection of pairwise disjoint nonempty sets and $\mathscr{D}$ is a subcollection such that $\cup\mathscr{D} = \cup\mathscr{C}$, then $\mathscr{D} = \mathscr{C}$.

4. Prove that (b) $\Rightarrow$ (c) in Theorem 7.7.

5. Why is it true that if C is an open set and some sequence $\{x_n\}$ converges to a point of C, then $\{x_n\}$ is ultimately in C?

8. CONTINUOUS FUNCTIONS

Let f: $D \to R$ be a function whose domain D is a subset of R. We say f is *continuous at a point* x of D provided that if $\{x_n\}$ is any sequence in D converging to x, then $\{f(x_n)\}$ converges to $f(x)$. If f is continuous at every point of D we say f is *continuous on* D, or just *continuous*.

The identity function, $f(x)=x$, is continuous (on any domain), and any constant function is continuous. The function f: $R \to R$ given by $f(x) = 0$ for $x \leq 0$ and $f(x) = 1$ for $x > 0$ is not continuous because $1/n \to 0$, but $\{f(1/n)\}$ is the constant sequence $\{1\}$ which does not converge to $f(0) = 0$.

Translating the theorem on algebraic properties of limits we get the following theorem on algebraic properties of continuous functions.

THEOREM 8.1 *Let* f, g: $D \to R$ *be continuous at the point* x *of* D. Then

(a) *The function* $f + g$: $D \to R$ *defined by* $(f+g)(x) = f(x) + g(x)$ *is continuous at* x.
(b) *The function* $f \cdot g$: $D \to R$ *defined by* $(f \cdot g)(x) = f(x) \cdot g(x)$ *is continuous at* x.
(c) *If* K *is a constant, the function* Kf: $D \to R$ *defined by* $(Kf)(x) = K \cdot f(x)$ *is continuous at* x.
(d) *If* $g(x) \neq 0$ *for all* x *in* D, *then the function* f/g: $D \to R$ *defined by* $(f/g)(x) = f(x)/g(x)$ *is continuous at* x.

Using this theorem we have immediately that the function $f(x) = x^2$ is continuous (any domain), since it is the product of the identity function with itself. More generally, any function of the form $f(x) = x^n (n \in I^+)$ is continuous, hence any polynomial function is continuous, and by part (d) any rational function (quotient of polynomials) is continuous on its domain.

Our definition of continuity was chosen to exploit the work we have done on sequential convergence. We now show that it is equivalent to the usual "$\varepsilon - \delta$" definition of continuity.

THEOREM 8.2 *Let $f: D \to R$ and let x be a point of D. Then f is continuous at x if and only if the following condition holds.*
($*$) *Given $\varepsilon > 0$, there is $\delta > 0$ such that if $y \in D$ and $|x - y| < \delta$, then $|f(y) - f(x)| < \varepsilon$.*

PROOF. If ($*$) fails, there is an $x \in D$ and an $\varepsilon > 0$ such that if δ is any positive number, then there is a point $y \in D$ with $|y - x| < \delta$ but $|f(y) - f(x)| \geq \varepsilon$. In particular, for each number of the form $1/n$ ($n \in I^+$), there is a point y_n of D with $|y_n - x| < 1/n$ but $|f(y_n) - f(x)| \geq \varepsilon$. Then $\{y_n\}$ is a sequence in D converging to x such that $\{f(y_n)\}$ does not converge to $f(x)$; that is, f is not continuous at x. We have proved that if f is continuous at x, then ($*$) holds.

To prove the converse, suppose ($*$) holds and let $\{x_n\}$ be a sequence in D converging to x; we wish to prove that $\{f(x_n)\}$ converges to $f(x)$. Suppose $\varepsilon > 0$ is given; let δ be as in ($*$) and choose $N \in I^+$ such that if $n \geq N$, then $|x_n - x| < \delta$. Then if $n \geq N$ we also have $|f(x_n) - f(x)| < \varepsilon$, as desired. ∎

In this section we are going to establish two important properties of continuous functions.

Before stating the first result let us review a familiar concept. Suppose $f: D \to R$ and the range of f is bounded above. The l.u.b., M, of the range is called the *maximum of f on D* and if there is a point x of D such that $f(x) = M$, we say *f achieves* its maximum on D (at x). Similar definitions hold for *minimum.*

Here are some examples. The function f: $(0, 1) \to R$ given by $f(x) = x^2$ has a maximum of 1 and a minimum of 0. It achieves its minimum but not its maximum. The function f: $[-1, 1] \to R$ defined by $f(x) = -1$ if $x \leq 0$ and $f(x) = 1/x$ if $x > 0$ has no maximum; it achieves its minimum, which is -1, at many points.

The following result tells when a function achieves its extreme values.

 THEOREM 8.3 *Let $f: D \to R$ be continuous and suppose D is* *compact. Then f has a maximum and minimum on D and achieves each of these.*

PROOF. Let E denote the range of f; $E = \{f(x) \mid x \in D\}$. We first show that E is compact by showing that it has property (c) of the theorem on compactness. Let $\{x_n\}$ be a sequence in E. For each n, let y_n be a point of D such that $f(y_n) = x_n$. Then $\{y_n\}$ is a sequence in D and since D is compact there is a subsequence $\{z_n\}$ which converges to a point y of D. Then $\{f(z_n)\}$ is a subsequence of $\{x_n\}$ (?) and converges to $f(y)$, which is in E.

We have shown that E is compact, hence E is closed and bounded. Let $M = \text{l.u.b.}E$ and $m = \text{g.l.b.}E$; then M is the maximum of f on D and m the minimum. Since E is closed it contains M; otherwise, there would be an interval (a, b) such that $a < M < b$ and $(a, b) \cap E = \varnothing$. It would then follow that M is not the least upper bound of E, a contradiction. Thus $M \in E$, which implies that, for some x in $D, f(x) = M$. Similarly, E contains m and hence achieves its minimum on D. ▮

To prove the final result of this section we will use the following result.

LEMMA 8.4 *Let f: $D \to R$ be continuous and let $x \in D$.*

(a) *If $f(x) > c$, then there is an open interval A such that $x \in A$ and if $y \in A \cap D$, then $f(y) > c$.*
(b) *If $f(x) < c$, then there is an open interval A such that $x \in A$ and if $y \in A \cap D$, then $f(y) < c$.*

PROOF. We prove (a); the proof of (b) is similar. Thus suppose f: $D \to R$ is continuous, $x \in D$ and $f(x) > c$. Let $\varepsilon = f(x) - c$ and choose $\delta > 0$ so that if $y \in D$ and $|x - y| < \delta$, then

$$|f(y) - f(x)| < \varepsilon.$$

Let $A = (x - \delta, x + \delta)$; then if $y \in A \cap D$, we have $|y - x| < \delta$ and hence $|f(y) - f(x)| < \varepsilon$. This last condition says that $f(y)$ is in $(f(x) - \varepsilon, f(x) + \varepsilon)$. Thus if $y \in A \cap D$, then $f(y) > f(x) - \varepsilon = c$. ▮

The next result can be interpreted as saying that there are no holes (or jumps) in the graph of a continuous function defined on an interval.

INTERMEDIATE VALUE THEOREM 8.5. *Let $f\colon D \to R$ be continuous where $D = [a, b]$. Suppose c is a number lying between $f(a)$ and $f(b)$. Then there is an element x of D such that $f(x) = c$.*

PROOF. If $c = f(a)$ or $c = f(b)$, the result is trivial. The other two possibilities are $f(a) < c < f(b)$ or $f(a) > c > f(b)$. For definiteness assume that $f(a) < c < f(b)$.

$$A = \{x \in D \mid c < f(x)\};$$

then A is a nonempty subset of R [it contains $f(b)$] and is bounded below (by c). Let $x = \text{g.l.b.}A$; then $a < x < b$ and we assert that $f(x) = c$.

First, suppose $f(x) > c$. By part (a) of Lemma 8.4 there is an interval (a', b') containing x such that for every y in $(a', b') \cap D$, $f(y) > c$. Let y be a point of $(a', b') \cap D$ which is less than x; then $f(y) > c$, hence $y \in A$. But this contradicts the fact that x is a lower bound for A. So, after all, we must have $f(x) \leq c$.

If $f(x)$ were less than c, then using part (b) of Lemma 8.4 and arguing as above, there would be a point y of D such that $x < y$ and y is a lower bound for A. This again contradicts the choice of x and shows that $f(x) \geq c$.

Thus we have $f(x) \leq c$ and $f(x) \geq c$ and hence $f(x) = c$, as asserted. ∎

EXERCISES

1. Let $f\colon D \to R$ and let $x \in D$. Show that f is continuous at x if and only if given any open interval A which contains $f(x)$, there is an open interval B containing x such that if $y \in B \cap D$, then $f(y) \in A$.

2. Show that $f\colon R \to R$ is continuous if and only if given an open set $O \subseteq R$, the set $\{x \in R \mid f(x) \in O\}$ is open.

3. Prove that the composition of continuous is continuous. (This has to be stated precisely first.)

4. Suppose you know that any interval $[a, b]$, where $a < b$, is uncountable. Prove that if $f\colon [a, b] \to R$ is continuous, then the range of f is either a single point or is uncountable.

5. Supply the missing details to show that $f(x) \geq c$ in the proof of the intermediate value theorem.

6. (a) Suppose X is a subset of R which is not bounded. Show there that is a function $f\colon X \to R$ which is continuous and has no maximum on X. (This is easy.)

(b) Do (a) with "not bounded" replaced by "not closed." (*Hint:* Let x be a limit point of X which is not in X. Make f misbehave near x.)

7. Prove the following fixed-point theorem: If $f\colon [0, 1] \to R$ is continuous and the range of f is a subset of $[0, 1]$, then there is a point x of $[0, 1]$ such that $f(x) = x$. (*Hint:* Apply the intermediate value theorem to the function $f = f - i$, where i is the identity function.)

8. Let f be a function from R into R. Which of the following imply continuity of f and which are implied by continuity of f?

(a) If $\{x_n\}$ is Cauchy, then $\{f(x_n)\}$ is Cauchy.
(b) If A is closed, then $\{x \mid f(x) \in A\}$ is closed.
(c) If A is open, then $\{x \mid f(x) \in A\}$ is open.
(d) If A is closed, then $\{f(x) \mid x \in A\}$ is closed.
(e) If A is open, then $\{f(x) \mid x \in A\}$ is open.
(f) If $\{x_n\}$ converges, then $\{f(x_n)\}$ has a convergent subsequence.
(g) If $\{x_n\}$ converges, then every subsequence of $\{f(x_n)\}$ converges.

CHAPTER 6

Infinite Series

INTRODUCTION

In almost every area of analysis infinite series play an important role. We give two illustrations, both of which will be encountered by the reader early in his mathematical training.

In advanced calculus it is shown that any "sufficiently nice" function can be expressed in a natural way as an infinite series: the Taylor's series of the function. The so-called *elementary* functions—the trigonometric, exponential, and logarithmic functions—are sufficiently nice and their properties are quite easily derived by investigating the corresponding series.

In differential equations, series are used to prove existence theorems, that is, theorems which guarantee that an equation has a solution. Moreover, sometimes the only "reasonable" way to write down the solution of a given differential equation is by means of an infinite series.

The theory of infinite series may be considered an application of the theory of sequences. There is a special notation and terminology used for series; once this has been introduced, much of the theory is obtained by translating previously known facts about sequences.

2. BASIC DEFINITIONS

We begin with the so-called "sigma notation." Let $\{a_i \mid i \in I^+\}$ be a sequence (of real numbers) and let m and n be positive integers with $m \leq n$. The sum of the mth through nth terms of the sequence, that is, the sum $a_m + a_{m+1} + \cdots + a_n$, will be denoted by

$$\sum_{i=m}^{n} a_i \qquad \text{or} \qquad \sum_{m}^{n} a_i \qquad \text{or} \qquad \textstyle\sum_m^n a_i.$$

Either of these may be read, "the sum of the a_i's as i runs from m to n." Here are some examples.

$$\sum_{1}^{2} a_i = a_1 + a_2,$$

$$\sum_{3}^{5} a_i = a_3 + a_4 + a_5,$$

$$\sum_{10}^{10} a_i = a_{10}.$$

We can now state the basic definitions of this chapter. An *infinite series* (of reals) is a symbol $\sum a_i$, where $\{a_i\}$ is a sequence (of reals). Frequently, we just say "series" instead of "infinite series." The nth *term* of the series $\sum a_i$ is a_n. Given a series $\sum a_i$, the nth *partial sum of the series* is the number $\sum_1^n a_i$. The *sequence of partial sums* of the series is the sequence whose nth term is the nth partial sum; thus the sequence of partial sums of $\sum a_i$ is the sequence $\{S_n\}$, where

$$S_n = \sum_1^n a_i .$$

Example 1. If $a_i = (-1)^i$, then the partial sums of $\sum a_i$ are $S_1 = -1$, $S_2 = 0$, $S_3 = -1$, and, in general, $S_n = -1$ if n is odd, 0 if n is even.

Example 2. The partial sums of $\sum 1/2^i$ are $S_1 = \frac{1}{2}$, $S_2 = \frac{3}{4}$ $S_3 = \frac{7}{8}$, and, in general, $S_n = (2^n - 1)/2^n$.

Example 3. The partial sums of the series corresponding to $a_i = i$ are $S_1 = 1$, $S_2 = 3$, $S_3 = 6$, and, as may be proved by induction, $S_n = 1 + 2 + \cdots + n = n(n+1)/2$.

We say that the series $\sum a_i$ *converges* and that its *sum* is A provided that the sequence of partial sums of the series converges to A. In this case we write $\sum a_i = A$. If the sequence of partial sums diverges (does not converge), we say that the series *diverges*. Thus a statement about convergence or divergence of a series is merely a statement about its sequence of partial sums.

Example 1. The series $\sum(-1)^i$ diverges (why?).

Example 2. Consider the series $\sum 1/2^i$; here the nth partial sum is $S_n = (2^n - 1)/2^n$. A trivial computation shows that $\{S_n\}$ converges to 1. Thus $\sum 1/2^i = 1$.

Example 3. The series $\sum i$ diverges because the sequence $\{n(n+1)/2\}$ diverges.

EXERCISES

1. Prove this theorem. If $\{a_i\}$ and $\{b_i\}$ are ultimately equal, say $a_i = b_i$ for all $i \geq N$, then $\sum a_i$ converges if and only if $\sum b_i$ converges.

2. What is a necessary and sufficient condition for convergence of the series $\sum a_i$ where, for all i, $a_i = K$ (a constant)?

3. Tell whether or not these series converge. If possible find a formula for the nth partial sum, S_n, of each.

(a) $\sum[(-1)^i - 1]$
(b) $\sum a_i$, where $a_i = 0$ if i is even, $2^{-(i+1)/2}$ if i is odd
(c) $\sum(-1)^i i$
(d) $\sum \frac{1}{3^i}$
(e) $\sum \frac{|\cos \alpha_i|}{2^i}$, where $\{\alpha_i\}$ is any sequence of numbers
(f) $\sum \frac{1}{n(n+1)}$
(g) $\sum \frac{1}{2n(2n+1)}$

4. Let x denote a real number. For what values of x does the series $\sum(n+1)x^n$ converge? (*Hint:* Try to express the nth partial sum in terms of n and x.)

5. Same as Exercise 4 for the series $\sum ne^{-nx}$.

3. ELEMENTARY FACTS

We now begin the process of translating old facts about sequences to new ones about series.

THEOREM 3.1 *If* $\sum a_i = A$ *and* $\sum b_i = B$, *then*

(a) $\sum (a_i + b_i) = A + B$.
(b) $\sum Ka_i = KA$ (K *a constant*).

PROOF. Let $\{S_n\}$ and $\{T_n\}$ be the sequences of partial sums of $\sum a_i$ and $\sum b_i$, respectively; the sequence $\{R_n\}$ of partial sums of $\sum (a_i + b_i)$ is then given by

$$\begin{aligned} R_n &= (a_1 + b_1) + (a_2 + b_2) + \cdots + (a_n + b_n) \\ &= (a_2 + a_2 + \cdots + a_n) + (b_1 + b_2 + \cdots + b_n) \\ &= S_n + T_n. \end{aligned}$$

By hypothesis $S_n \to A$ and $T_n \to B$; thus $R_n \to A+B$ (why?). This proves (a) and a similar argument proves (b). ∎

Using the Cauchy criterion for convergence of a sequence we next derive a very useful divergence criterion for series, called the *n*th-term test.

THEOREM 3.2 *If* $\sum a_i$ *converges, then* $a_i \to 0$.

We remark that it is the contrapositive of this statement which is quoted most often: *If* $a_i \nrightarrow 0$, *then* $\sum a_i$ *diverges.*

The theorem is proved as follows. As usual, let S_n denote the nth partial sum of $\sum a_i$. If the series converges, then, by definition, $\{S_n\}$ converges and is Cauchy. Given $\varepsilon > 0$, choose $N \in I^+$ so that if $m,n \geq N$, then $|S_m - S_n| < \varepsilon$. Then, in particular, if $i \geq N+1$, we have $|a_i| = |S_i - S_{i-1}| < \varepsilon$. Thus for every $\varepsilon > 0$, $|a_i|$ is ultimately less than ε, which says that $a_i \to 0$. ∎

If one is asked to determine whether a given series converges or diverges, the first order of business should be to test for divergence by using the nth term test. For example, this test shows immediately that the series $\sum (-1)^i$ and $\sum i$ diverge. We caution the reader that the converse of this theorem, "If $a_i \to 0$, then $\sum a_i$ converges," is *false*. Thus the nth-term test can never be used to determine convergence.

Here is a famous example, the *harmonic series*, which illustrates this fact. The harmonic series is the series $\sum 1/i$. Certainly $1/i \to 0$. Let S_n be the nth partial sum of this series. Then for any $m \in I^+$ we have

$$\begin{aligned} |S_{2m} - S_m| &= \frac{1}{2m} + \frac{1}{2m-1} + \cdots + \frac{1}{m+2} + \frac{1}{m+1} \\ &> \frac{1}{2m} + \frac{1}{2m} + \cdots + \frac{1}{2m} + \frac{1}{2m} (m \text{ terms}) \\ &= \tfrac{1}{2}. \end{aligned}$$

In particular, $\{S_n\}$ is not Cauchy (why?); hence it and the series diverge.

We close this section by stating a result which completely solves the problem of convergence for one particular class of series. A *geometric series* is one of the form $\sum ar^{i-1}$, where a and r are nonzero real numbers. For such series we have the following

THEOREM 3.3 $\sum ar^{i-1}$ *converges if and only if* $|r| < 1$, *in which case its sum is* $a/1 - r$.

PROOF. The "only if" part of the theorem follows from the nth-term test. If $|r| \geq 1$, then $\{r^{i-1}\}$ does not converge to 0; hence, since $a \neq 0$, neither does $\{ar^{i-1}\}$ and the series diverges.

The nth partial sum of $\sum ar^{i-1}$ is given by

$$\begin{aligned} S_n &= a + ar + \cdots + ar^{n-1} \\ &= a(1 + r + \cdots + r^{n-1}) \\ &= a\left(\frac{1 - r^n}{1 - r}\right) \text{ (why?).} \end{aligned}$$

A simple computation then shows that $S_n \to a/1 - r$ if $|r| < 1$. ∎

The number r which appears in a geometric series is called the *common ratio* of the series (for obvious reasons). The convergence of a geometric series is completely determined by the absolute value of its common ratio, so it is useful to be able to recognize a geometric series quickly. This problem is discussed in the exercises.

EXERCISES

1. Show that if $\sum a_i = A$ and K is a constant, then $\sum Ka_i = KA$. Prove or disprove that if $\sum a_i = A$ and $\sum b_i = B$, then $\sum(a_i b_i) = AB$.

2. Let $\{a_i\}$ be any sequence of nonzero real numbers. Show that $\sum a_i$ can be written as a geometric series with common ratio r if and only if $a_{i+1}/a_i = r$ for every $i \in I^+$. (*Note:* If $\sum a_i$ is of the form $\sum ar^{i-1}$, then we must have $a_1 = a$.)

3. Which of the following are geometric series? For those which are, tell what a and r are.

   (a) $\sum \dfrac{(-1)^{i-1}5^{i-1}}{3^{i-1}}$ (b) $\sum \dfrac{7^{i+2}2^{i-1}}{3^i}$

   (c) $\sum \dfrac{2^{3i}}{10^5}$ (d) $\sum \dfrac{2^{i-1}i^2}{7^i}$

(e) $\sum \frac{\cos \pi i}{7^{i+2}}$ (f) $\sum \frac{i+2}{i+3}$

(g) $\sum \frac{3^{i/2}}{4^i}$ (h) $\sum \frac{3^i}{4^i}$

4. Let A be the subset of $R \times R$ indicated in Figure 6.1. What is the area of that part of A lying over the interval $[0, 1]$? $[0, 3]$? What would you say the "total area" of A is and why?

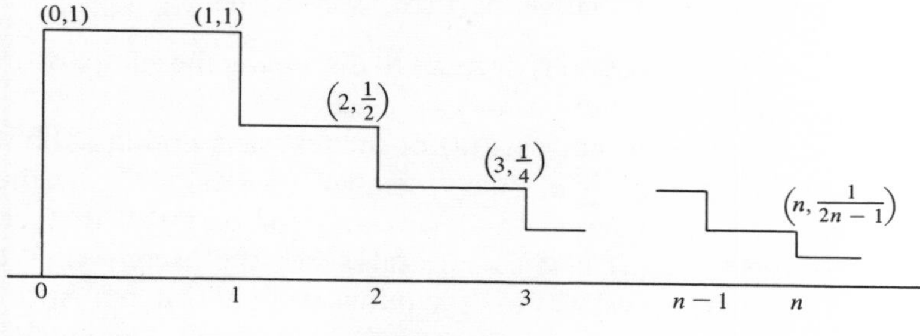

FIGURE 6.1

4. SERIES WITH NONNEGATIVE TERMS

For a while we are going to be concerned with finding convergence (or divergence) tests for series. Quite often these tests take the form of conditions on the terms of the series, an example being the nth-term test for divergence. In this section we shall consider series all of whose terms are nonnegative, that is, series $\sum a_i$, where, for each i, $a_i \geq 0$.

The first result gives a necessary and sufficient condition for the convergence of such series.

THEOREM 4.1 *A series with nonnegative terms converges if and only if its sequence of partial sums* $\{S_n\}$ *is bounded, in which case the sum of the series is* l.u.b. $\{S_n\}$.

PROOF. If a series has nonnegative terms, then its sequence of partial sums is nondecreasing. Such a sequence converges if and only if it is bounded and, in this case, it converges to its l.u.b. ▮

Here is the important *comparison test* for convergence.

THEOREM 4.2 *Let* $\{a_i\}$ *and* $\{b_i\}$ *be sequences satisfying* $0 \leq a_i \leq b_i$, *for all i. Then*

(a) *If* $\sum b_i$ *converges, so does* $\sum a_i$.

(b) *If* $\sum a_i$ *diverges, so does* $\sum b_i$.

PROOF. Let $\{S_n\}$ and $\{T_n\}$ be the sequences of partial sums of $\sum a_i$ and $\sum b_i$, respectively. Each sequence is nondecreasing and, for each n, $S_n \leq T_n$. If $\sum b_i$ converges, then, by Theorem 4.1, $\{T_n\}$ is bounded. But then $\{S_n\}$ is also bounded and, again, by Theorem 4.1, $\sum a_i$ converges. Thus part (a) is established and therefore its contrapositive, part (b), is also true. ▮

The rest of this section is devoted to discussing the comparison test and illustrating how it may be used.

First, we observe that part (a) of the test says nothing about the convergence of $\sum a_i$ (the "smaller" series) if $\sum b_i$ (the "larger" series) diverges, and similarly for part (b). In fact, the converses of (a) and (b) are false (see the exercises). We suggest that the easiest way to keep things straight is to remember the rather trivial proof of theorem, which simply amounts to comparing two nondecreasing sequences.

Quite frequently it takes some ingenuity to apply the comparison test. Suppose we are given a series $\sum a_i$ with nonnegative terms and we suspect (somehow) that it converges. To use part (a) of the theorem we must find a series $\sum b_i$ which we know is convergent and which satisfies $a_i \leq b_i$, for all i.

To illustrate this, consider the series $\sum 1/i^i$. The first few partial sums are 1, $\frac{5}{4}$, $\frac{139}{118}$, but, clearly, it would be difficult to find a formula for the nth partial sum from which convergence could be determined directly. However, notice that, for every $i \in I^+$, we have $1/i^i \leq 1/2^{i-1}$ (verify this). The geometric series $\sum 1/2^{i-1}$ converges so, by part (a), the series $\sum 1/i^i$ converges.

A similar process is involved when part (b) is to be used. Suppose we are asked to investigate the series $\sum 1/\sqrt{i}$. This is obviously related to the harmonic series $\sum 1/i$. In fact, we have $1/i \leq 1/\sqrt{i}$, for all $i \in I^+$, and, since the harmonic series diverges, part (b) of the comparison test says that the given series also diverges.

EXERCISES

1. Show that the comparison test is still true if we replace the words "for all i" by the words "for all but finitely many i."

2. Show that the converse of part (a) of the comparison test is false in general.

3. A series $\sum a_i$ has nonpositive terms if $a_i \leq 0$, for all i. State and prove the analogues of the two theorems of this section for such series. (There is a short way of doing this.)

4. A series $\sum b_i$ is a *rearrangement* of a series $\sum a_i$ if there is a 1-1 function $f: I^+ \to I^+$ such that $b_i = a_{f(i)}$, for all i (thus the terms of the series are the same but may occur in different order). Show that *if $\sum a_i$ has nonnegative terms and is convergent, $\sum a_i = A$, then every rearrangement converges and has sum A.* [*Hint:* Suppose $\sum b_i$ is a rearrangement of $\sum a_i$. Show that A is an upper bound for the partial sums of $\sum b_i$ and apply the first theorem of this section. Now use the fact that $\sum a_i$ is a rearrangement of $\sum b_i$ (why?).

5. The series $\sum_{n=1}^{\infty} 1/n^p$, where p is a real number, is called the *harmonic series of order p*. Show that it diverges if $p \leq 1$ and converges otherwise.

6. Suppose $\{a_i\}$ is a decreasing sequence of positive real numbers. For each n, let $b_n = 2^n a_{2n}$. Prove that $\sum a_n$ and $\sum b_n$ both converge or diverge.

7. If $\sum a_n$ is a convergent series with positive terms, prove that the series $\sum \sqrt{a_n a_{n+1}}$ and $\sum 2a_n a_{n+1}/a_n + a_{n+1}$ converge.

5. INTEGRAL TEST

From Exercise 5, Section 4, we see that the harmonic series of order p, $\sum 1/n^p$ converges if $1 < p$ and diverges otherwise. If we consider the function f_p defined on the set of positive real numbers by $f_p(x) = x^{-p}$, we see that the nth term of this series is just $f_p(n)$.

We now observe that the series $\sum 1/n^p$ and the improper integral $\int_1^{\infty} f_p(x)\,dx$ behave the same way. More precisely, one converges if and only if the other does. To see this suppose first that $p = 1$; then $\int_1^{\infty} f_p(x)\,dx = \int_1^{\infty} dx/x = \lim_{N\to\infty} \int_1^N dx/x = \lim_{N\to\infty} (\ln N) = \infty$. If $p \neq 1$, then

$$\int_1^{\infty} f_p(x)\,dx = \lim_{N\to\infty} [(N^{1-p} - 1)/1 - p],$$

which is ∞ if $1 - p > 0$ and $1/1 - p$ if $1 - p < 0$. To summarize, if $p \leq 1$, then $\int_1^{\infty} f_p(x)\,dx$ and $\sum 1/n^p$ diverge, and if $p > 1$, then both converge.

This connection between a series and an integral illustrates the following result.

THEOREM 5.1 **The Integral Test** *Let f be a function defined on the positive reals which is continuous, nonincreasing and positive valued. The series $\sum f(n)$ converges if and only if the improper integral $\int_1^\infty f(x)\,dx$ converges.*

PROOF. The conditions on f imply that for each positive integer n we have $f(2)+f(3)+\cdots+f(n) \leq \int_1^n f(x)\,dx \leq f(1)+f(2)+\cdots+f(n-1)$. It follows that if $\int_1^\infty f(x)\,dx$ converges, say $\int_1^\infty f(x)\,dx = A$, then every partial sum of $\sum f(n)$ is bounded above by $A+f(1)$, hence this series also converges. Conversely, if $\sum f(n)$ converges, say with sum B, then B is an upper bound for the sequence $\{\int_1^n f(x)\,dx\}$. Thus this sequence converges; that is, $\int_1^\infty f(x)\,dx$ converges. ▮

Example Consider the series $\sum 1/(n+1)\ln(n+1)$. The function $f(x) = 1/(x+1)\ln(x+1)$ satisfies the hypotheses of the integral test and we have $\int_1^n f(x)\,dx = \ln(\ln(x+1))|_1^n = \ln\ln(n+1) - \ln\ln 2$. This sequence diverges to ∞; hence the series we are considering also diverges.

6. SERIES WITH POSITIVE AND NEGATIVE TERMS

Except for the case of geometric series, the tests given so far apply only to series with nonnegative terms. Series without this property may sometimes be handled by the methods of this section.

Let $\sum a_n$ be any series. We say that $\sum a_n$ is *absolutely convergent* provided that the associated series $\sum |a_n|$ is convergent. If this terminology is to make sense, then an absolutely convergent series should be convergent; this is indeed so, but the proof is not trivial.

Suppose $\sum a_n$ is absolutely convergent. Form a new series, $\sum c_n$, whose nth term is given by $c_n = a_n + |a_n|$. We then have $0 \leq c_n \leq 2|a_n|$ for each n. It follows that $\sum c_n$ converges (why?). Since $\sum a_n = \sum (c_n - |a_n|)$ and since $\sum c_n$ and $\sum |a_n|$ both converge, we know that $\sum a_n$ itself converges.

We now give two important tests for absolute convergence. These also apply to series with nonnegative terms.

THEOREM 6.1 **The Ratio Test** *Let $\sum a_n$ be a series with nonzero terms and suppose the sequence of ratios $\{|a_{n+1}/a_n|\}$ converges, say to L. Then*

(a) *$\sum a_n$ is absolutely convergent if $L < 1$.*
(b) *$\sum a_n$ is divergent if $L > 1$.*
(c) *If $L = 1$, $\sum a_n$ may converge or diverge.*

PROOF.

(a) Suppose $|a_{n+1}/a_n| \to L < 1$. Let r be any number satisfying $L < r < 1$. Choose N sufficiently large so that if $n \geq N$, then $|a_{n+1}/a_n| < r$. We then have

$$|a_{N+1}| < r|a_N|,$$
$$|a_{N+2}| < r\,|a_{N+1}|,$$

and, in general,

$$|a_{N+m}| < r^m\,|a_N|.$$

Thus all but finitely many terms of $\sum |a_n|$ are less than the corresponding terms of the geometric series $\sum ar^n$, where $a = |a_N|$. This geometric series converges (why?), and therefore so does $\sum |a_n|$.

(b) If $L > 1$, then there is an integer N such that if $n \geq N$, then $|a_N| < |a_n|$ (why?). Since $0 < |a_N|$, the sequence $\{a_n\}$ cannot converge to 0; hence $\sum a_n$ diverges.

(c) To prove part (c) one must exhibit two series satisfying the hypotheses of the theorem and for which $L = 1$ such that one of the series diverges and the other converges. This is left as an exercise for the reader with the suggestion that he consider harmonic series of various orders. ∎

THEOREM 6.2 **The Root Test** *Given the series $\sum a_n$, suppose the sequence whose* n*th term is $\sqrt[n]{|a_n|}$ converges to L. Then*

(a) *If $L < 1$, $\sum a_n$ is absolutely convergent.*

(b) *If* $L > 1$, $\sum a_n$ *diverges.*

(c) *If* $L = 1$, $\sum a_n$ *may diverge or converge.*

We sketch a proof of parts (a) and (b). If $L < 1$, then there is a number $r < 1$ such that for all but finitely many n, $\sqrt[n]{|a_n|} < r$; that is, $|a_n| < r^n$. Now compare $\sum |a_n|$ with $\sum r^n$. If $L > 1$, then for all but finitely many n, we have $1 < |a_n|$; hence by the nth root test $\sum a_n$ cannot converge. ▮

Absolute convergence is a lot to ask of a series. It may happen that $\sum a_n$ is convergent but $\sum |a_n|$ is not. In particular, this can happen when $L = 1$ in either the ratio test or the root test. A more delicate approach yields an elegant convergence criterion for certain types of series: series in which successive terms have opposite signs.

An *alternating series* is one of the form $\sum (-1)^{n+1} a_n$, where $a_n \geq 0$ for all n.

THEOREM 6.3 **Alternating Series Test** *Let* $\sum (-1)^{n+1} a_n$ *be an alternating series and suppose that for each* n, $a_{n+1} \leq a_n$; *then* $\sum (-1)^{n+1} a_n$ *converges if and only if* $a_n \to 0$.

PROOF. Suppose $\sum (-1)^{n+1} a_n$ satisfies the hypotheses of the theorem and that $a_n \to 0$. Let S_n denote the nth partial sum of this series. Consider the sequence $\{S_{2n}\}$ of partial sums with even subscripts. We assert that $\{S_{2n}\}$ is nondecreasing and bounded above. The first part of this statement follows from the inequality $S_{2n+2} - S_{2n} = a_{2n+1} - a_{2n+2} \geq 0$ and the second from $S_{2n} = a_1 + (-a_2 + a_3) + \cdots + (-a_{2n-2} + a_{2n-1}) - a_{2n} < a_1$. Hence $\{S_{2n}\}$ converges to its l.u.b., say l.

Similar computations show that the sequence of partial sums with odd subscripts is nonincreasing and bounded below, hence converges to its g.l.b., L.

Since $S_{2n} - S_{2n-1} = a_{2n}$ and since the sequence $\{a_n\}$ converges to 0, it follows that $l = L$. It is then easy to see that $\{S_n\}$ also converges to this common limit. This proves the "if" part of the theorem.

The "only if" part is just the nth-term test. If $\sum (-1)^{n+1} a_n$ converges, then $\{(-1)^{n+1} a_n\}$ converges to 0, hence $\{a_n\}$ converges to 0. ▮

The theorem we have just proved may be interpreted as saying that if we have a series *which satisfies the given hypotheses* and if convergence is not automatically ruled out by the nth-term test, then, in fact, the series converges.

As an illustration consider the series $\sum (-1)^{n+1}/n$. It is easily verified that none of the tests previously established applies to this series. In particular this series is not absolutely convergent, because upon taking absolute values we get the harmonic series which diverges. However, the alternating series test shows that this series converges.

In this chapter we have given only a few of the many tests for convergence and divergence; these are the standard and most applied ones. There is a large number of other tests, many of which are refined versions of the ones exhibited, some very different in their character.

EXERCISES

In the following problems you are given the nth term, a_n, of a sequence. Discuss the convergence or divergence of $\sum a_n$ using any test in this chapter.

1. $a_n = (1 + a^n)^{-1}$, a a positive constant

2. $a_n = n!\, n^{-n^2}$

3. $a_n = ne^{-n^2}$

4. $a_n = n^{-1/2}(n + 1)^{-1/2}$

5. $a_n = \dfrac{(n!)^2}{(2n)!}$

6. $a_n = (\ln n)^{-\ln \ln n}$

7. $a_n = \cos (n^2\pi)$

8. $a_n = \dfrac{1}{n \ln n(\ln \ln n)^p}$, $p > 1$; $p < 1$; $p = 1$; $n \geq 100$.

9. $a_n = \left(\dfrac{n}{n+1}\right)^{n^2}$

10. $a_n = \dfrac{(-1)^n}{n!}$

11. $a_n = \dfrac{1}{(n + 1)(n + 2)}$

12. $a_n = \dfrac{n^n}{n!}$

13. $a_n = \dfrac{n + 1}{\ln (n + 2)}$

14. $a_n = \dfrac{3^{n+2}}{n^3}$

15. $a_n = \dfrac{\ln n}{n + \ln n}$

16. $a_n = \dfrac{n^2}{n! + 1}$

17. $a_n = \dfrac{1 + (\ln n)^2}{n(\ln n)^2}$, $n \geq 2$

18. $a_n = \dfrac{(-1)^n}{\ln n}$, $n \geq 2$

19. $a_n = \dfrac{(-1)^n \ln n}{4n + 5}$

20. $a_n = \dfrac{1}{\sqrt{n} \ln n}$, $n \geq 2$

21. $a_n = \dfrac{1}{n^n}$

22. $a_n = \dfrac{2^n + 1}{3^n + 1}$

23. $a_n = \dfrac{n!}{3 \cdot 5 \cdot 7 \cdot \ldots \cdot (2n + 3)}$

24. $a_n = \dfrac{n^2}{e^n}$

25. $a_n = \dfrac{\sqrt{n}}{n^2 - 4}$, $n \geq 3$

26. Let $\sum a_n$ and $\sum b_n$ be two series with sequences of partial sums $\{A_n\}$ and $\{B_n\}$, respectively. Let $\{C_n\}$ be the partial sums for $\sum (a_n b_n)$.

(a) Verify

$$C_n - C_m = -b_m A_{m-1} + b_n A_n + \textstyle\sum_{k=m}^{n} (b_k - b_{k+1}) A_k .$$

(b) If (i) $\{A_n\}$ is bounded, (ii) $\lim b_n = 0$, and (iii) $\sum |b_n - b_{n+1}|$ converges, prove that $\sum a_n b_n$ converges.

27. Suppose that the sequence $\sum a_n$ with positive terms converges. For each n, let $R_n = \sum_{i=m+1}^{\infty} a_i$ and let $b_n = \sqrt{R_{n-1}} - \sqrt{R_n}$. Show that (a) the sequence $\{a_n/b_n\}$ converges to 0; (b) $\sum b_n$ converges.

28. Suppose the sequence $\sum d_n$ with positive terms diverges. For each n, let $D_n = \sum_{i=1}^{n} d_i$ and let $e_n = \sqrt{D_n} - \sqrt{D_{n-1}}$. Show that (a) the sequence $\{e_n/d_n\}$ converges to 0; (b) $\sum e_n$ diverges.

29. Show that the sequence whose nth term is $1 + \frac{1}{2} + \cdots + 1/n - \log(n + 1)$ converges to a number γ and that $\gamma < 1$. This number is called Euler's constant; it is not known whether γ is rational or not. The first few terms of the decimal expansion of γ are .577215 . . . .

30. *Decimal expansions.* Let a be a real number and define a sequence $a_0, a_1, a_2, \ldots$ as follows:

a_0 is the largest integer such that $a_0 \leq a$,

a_1 is the largest integer such that $a_0 + \dfrac{a_1}{10} \leq a$.

In general, a_n is the largest integer such that

$$a_0 + \frac{a_1}{10} + \cdots + \frac{a_n}{10^n} \leq a.$$

Note that for $i \geq 1$, $0 \leq a_i \leq 9$.

(a) Show that with the a_i's defined as above, $\sum_{i=0}^{\infty} a_i/10^i$ converges and has sum a. Call this the *decimal series* for a. The *decimal expansion* of a is the symbol $a_0 . a_1 a_2 \cdots$. This is shorthand for the decimal series.

(b) Show that if $\sum_{i=0}^{\infty} b_i/10^i$ is a series such that $0 \leq b_i \leq 9$ for all $i \geq 1$, then the series converges. Is it true that this series is the decimal series for its sum? (*Hint*: Consider $b_0 = 0$, $b_1 = 2$, $b_i = 9$ for $i \geq 2$.)

(c) We say the decimal expansion for *a repeats* provided there exist integers $n \geq 0$ and $k \geq 1$ such that for all $i \geq n$ we have $a_i = a_{i+k} = a_{i+2k} = a_{i+3k} = \cdots$. For example, the decimal expansion 1.562314 2314 2314 ... repeats with $n = 3$ and $k = 4$. Show that the decimal expansion for a repeats if and only if $a = m/n$, where m and n are integers. (*Hint:* If the expansion repeats and n and k are as above, consider the number $10^{n+k}a - 10^n a$.)

7. THE REALS ARE UNCOUNTABLE

In this section, we prove the theorem stated as the title. It will be assumed that the reader has worked through Exercise 30 of Section 6.

Let S be the set of all functions s from I^+ into $\{0, 1, 2, \ldots, 9\}$ which have the property (*): for every integer n there is an integer $m > n$, such that $s(n) \neq 9$. Given a number a in the interval $[0, 1)$, let s_a be the function defined by $s_a(n) = a_n$ where $.a_1 a_2 a_3 \cdots$ is the decimal expansion of a. The reader may check that, by definition, no decimal expansion terminates in nines (for example, the decimal expansion of $\frac{1}{2}$ is .500 ... and not .499...) and therefore, each s_a satisfies condition (*) and is an element of S. In fact, the function F: $[0, 1) \to S$ given by $F(a) = s_a$ is 1-1 and onto.

To prove that the set of reals is uncountable, it suffices to prove that S is uncountable. Supposing to the contrary that S is countable, let h be a 1-1 function from I^+ onto S. Thus for each n in I^+, $h(n)$ is a function in S. We now construct an element of S

which is not the image under h of any integer. Namely, let s be the function whose value at n is 0 if $(h(n))(n) \neq 0$ and 1 if $(h(n))(n) = 0$. Certainly s is an element of S and, for each integer n, $h(n) \neq s$ since $h(n)$ and s assign different values to the integer n. Thus there is no 1-1 function from I^+ onto S and S is uncountable.

CHAPTER 7

Power Series

1. BASIC DEFINITIONS

By a *power series in x* we mean a series of the form

$$\sum_{i=0}^{\infty} a_i x^i.$$

The nth partial sum of a power series is a polynomial of degree n in the variable x; the first few of these are $S_0(x) = a_0 x^0 = a_0$, $S_1(x) = a_0 + a_1 x$, $S_2(x) = a_0 + a_1 x + a_2 x^2$, and so on. Note that as a matter of convenience we let the index i begin with 0 rather than 1. From now on we designate the above series simply by $\sum a_i x^i$.

If we select a particular value for x, that is, take x equal to some real number, then the power series becomes a series of real numbers. For a given power series, the series corresponding to different values of x may have different convergence properties. For example, the power series $\sum (1/n+1)x^n$ converges when $x = 0$ or $x = -1$ and diverges when $x = 1$.

The rather surprising fact is that the set of values for which a power series converges always consists of some type of interval on the real line, where for the sake of simplicity we consider a single point as an interval. To show this we need the following important fact.

LEMMA 1.1 *If the power series $\sum a_i x^i$ converges for $x = x_1$ and if $|x_2| < |x_1|$, then the series converges absolutely for $x = x_2$.*

Before proving the lemma we observe that we may not have absolute convergence at $x = x_1$, as is illustrated by taking $x_1 = -1$ in the series $\sum (1/n+1)x^n$.

To prove the lemma, suppose $\sum a_n x_1^n$ converges and $|x_2| < |x_1|$. By the nth-term test, $\{a_n x_1^n\}$ converges to 0, hence so does the sequence $\{|a_n x_1^n|\}$. Choose a positive constant K such that for every n, $|a_n x_1^n| < K$. Then, for every n, we also have $|a_n x_2^n| = |a_n x_1^n| \cdot |x_2/x_1|^n \leq Kr^n$, where $r = |x_2/x_1| < 1$. It follows that $\sum |a_n x_2^n|$ converges by comparison with $\sum Kr^n$. ∎

THEOREM 1.2 *Let S be the set of real numbers x for which the power series $\sum a_i x^i$ converges. Then S is either the set $\{0\}$, the* *entire real line, or a bounded interval of one of the forms* $(-R, R)$, $[-R, R]$, $[-R, R)$, and $(-R, R]$, *where* R *is a positive real number.*

We begin the proof of the theorem by noting that $\sum a_i x^i$ must converge when $x = 0$. If it converges for no other value of x, then $S = \{0\}$. Suppose then that $x_1 \in S$, where $x_1 \neq 0$. Then by the lemma the interval $(-|x_1, |x_1|)$ is contained in S also. It follows that, if S is a bounded set, then it is of one of the four types of bounded intervals and, if S is unbounded, then it is the entire real line. [*Note:* We also get, for free, from the lemma that if S is the entire real line, then the series converges absolutely for every value of x, while if S is one the intervals listed, then the series converges absolutely on $(-R, R)$ (and possibly at one or both of the end points).] ▮

The set S of values for which a power series converges is called the *interval* (or the *region*) *of convergence*. If S is one of the four types of bounded intervals listed in the theorem, we say that R is the *radius of convergence*; if S is the entire line we sometimes say the *radius of convergence is infinite*.

The behavior of a power series at the end points of its interval of convergence must be checked. As we have noted, convergence inside the interval is absolute.

At this point let us remark that we might also consider series of the form $\sum a_i(x - x_0)^i$, where x_0 is some real number. This is called a *power series in* $x - x_0$. The theory of such series exactly parallels the theory of power series in x. For example, the interval of convergence is defined as before and is either $\{x_0\}$, the entire real line, or a bounded interval centered at x_0. In the following sections we will generally consider power series in x; however, it should be understood that all these results (suitably modified) go through for series in $x - x_0$.

Let us now give some examples.

Example 1. Consider the series $\sum x^n$. For a fixed value of x this is a geometric series with common ratio x. Hence it converges if and only if $|x| < 1$, that is, the interval of convergence is $(-1, 1)$.

Example 2. Consider the series $\sum x^n/n^n$. Fix any value of x and choose n_0 large enough so that $|x| < n_0$. Then, for $n \geq n_0$, we have $|x/n| < |x/n_0| < 1$. Hence $\sum_{i=n_0}^{\infty} x^n/n^n$ converges absolutely (by comparison with what?) and, in particular $\sum x^n/n^n$ converges. Thus the given series converges for all values of x and has the whole real line as interval of convergence.

Example 3. An argument similar to the one just given shows that the series $\sum n^n x^n$ converges only when $x = 0$; hence the interval of convergence is $\{0\}$.

Example 1′. The series $\sum (x - x_0)^n$ has $(x_0 - 1, x_0 + 1)$ as its interval of convergence.

Example 2′. The series $\sum (x - x_0)^n/n^n$ converges for all x.

Example 3′. The interval of convergence of $\sum n^n(x - x_0)^n$ is $\{x_0\}$.

2. DETERMINATION OF RADIUS OF CONVERGENCE

Two usually successful ways of finding the radius of convergence are consequences of the ratio and root tests of Chaper 6.

THEOREM 2.1 *Given the series $\sum a_n x^n$, suppose either that*

(a) *the sequence $\{|a_{n+1}/a_n|\}$ converges to ρ, or*
(b) *the sequence $\{\sqrt[n]{|a_n|}\}$ converges to ρ.*

If $\rho = 0$, the radius of convergence is infinite. Otherwise, the radius of convergence is $1/\rho$.

PROOF. Suppose (a) holds with $\rho \neq 0$. If x_1 is any number such that $|x_1| < 1/\rho$, then we have

$$\left|\frac{a_{n+1} x_1^{n+1}}{a_n x_1^n}\right| = |x_1| \left|\frac{a_{n+1}}{a_n}\right| \to |x_1| \cdot \rho < 1.$$

By the ratio test the series $\sum a_n x_1^n$ converges (absolutely).

A similar computation shows that if $|x_1| > 1/\rho$, then $\sum a_n x_1^n$ diverges. This proves that $1/\rho$ is the radius of convergence if $\rho \neq 0$.

The case that $\rho = 0$ is even easier and is left as Exercise 4. So is the proof in case we assume that condition (b) holds. ▮

EXERCISES

1. (a) Show that if $\{c_i\}$ is a sequence of positive constants and $\lim \sqrt[n]{c_n} = 1$, then $\sum a_n x^n$ and $\sum c_n a_n x^n$ have the same radius of convergence.

(b) Show that the two series $\sum a_n x^n$ and $\sum n a_n x^{n-1}$ have the same radius of convergence.

2. In each of the following you are given the general coefficient a_n of a power series $\sum a_n x^n$. Find the radius of convergence of each series. Also, where applicable, tell whether the series diverges or converges at the end points of the interval of convergence.

(a) $a_n = \dfrac{(-1)^n}{(n+1)^2}$ (b) $a_n = \dfrac{1}{\ln(n+2)}$

(c) $a_n = \dfrac{1}{n^2+1}$ (d) $a_n = n!$

(e) $a_n = \left(\dfrac{n}{n+1}\right)^{2n}$ (f) $a_n = n^2 e^{-n}$

(g) $a_n = \dfrac{(2n)!}{(n!)^2}$ (h) $a_n = (\ln n)^{-n} (n \geq 2)$

$a_0 = a_1 = 1$

(i) $a_n = \dfrac{(-1)^n}{n!} \dfrac{(n)^n}{e}$ (j) $a_n = (2^n + 3^n)^{-1}$

(k) $a_n = \left(1 + \dfrac{1}{2} + \cdots + \dfrac{1}{n}\right)(n \geq 1)$

$a_0 = 1$

(l) $a_n = \left(1 + \dfrac{1}{n}\right) -n^2 (n \geq 1)$

$a_0 = 1$

3. For what values of x does $\sum x^n/(1-x)^n$ converge?

4. Show that if $\rho = 0$ in Theorem 2.1, then the radius of convergence is infinite.

3. TAYLOR POLYNOMIALS AND TAYLOR'S FORMULA

Let S be the interval of convergence of a power series $\sum a_i x^i$; then the formula $f(x) = \sum a_i x^i$ defines a real-valued function on S.

We shall eventually show that inside the interval of converge a function defined this way by a power series is a "nice" function. More precisely it is differentiable (hence continuous) and the derivative is obtained by termwise differentiation of its power series. This section is devoted to establishing preliminary results.

The first fact which we need may be considered an extension of the mean value theorem of differential calculus.

THEOREM 3.1 **Cauchy's Theorem** *Let f and g be two functions which are continuous on $[a, b]$ and differentiable on (a, b) and suppose that for every x in (a, b), $g'(x) \neq 0$. Then there is a point $\bar{x}$ in (a, b) such that*

$$\frac{f(b)-f(a)}{g(b)-g(a)} = \frac{f'(\bar{x})}{g'(\bar{x})}.$$

PROOF. First note that by the mean value theorem, $g(b)-g(a) \neq 0$, so that the formula makes sense. Consider the function h defined on $[a, b]$ by

$$h(x) = [f(b)-f(a)]g(x) - [g(b)-g(a)]f(x).$$

Clearly h is continuous on $[a, b]$, differentiable on (a, b), and $h(a) = h(b)$. By the mean value theorem there is a point $\bar{x}$ in $[a, b]$ such that $h'(\bar{x}) = 0$; this yields the desired formula. ∎

Let f be a function defined on $[a, b]$ such that the first n derivatives of f exist at some point y in (a, b). The *nth-degree Taylor polynomial of f at y is the polynomial function*

$$P_n(x; y)$$

$$= f(y) + f'(y)(x-y) + f''(y)\frac{(x-y)^2}{2!} + \cdots + f^{(n)}(y)\frac{(x-y)^n}{n!}$$

Note that P_n depends on f.

We remark that for a fixed value of x, $P_n(x; y)$ may be considered as a function of y. If f has $n+1$ derivatives in (a, b), then this function of y is differentiable; indeed, upon differentiation and simplification one finds that, for fixed x, the derivative of $P_n(x; y)$ with respect to y is given by

$$P_n'(x; y) = f^{(n+1)}(y)\frac{(x-y)^n}{n!}.$$

We now state a result which says that if f is reasonably nice on $[a, b]$, then f differs from its Taylor polynomial by a remainder term which is of a prescribed form. More precisely, we have $f(x) = P_n(x; c) + R_n(x; c)$, where c is a point of (a, b). The function $R_n(x; c)$ is called the nth *remainder* (at c). In general, $R_n(x; c)$ can be expressed in various ways. We shall give what is called the *Lagrange's form* of the remainder.

THEOREM 3.2 **Taylor's Formula** *Suppose $f, f^1, \ldots, f^{(n)}$ are continuous on $[a, b]$ and $f^{(n+1)}$ exists on (a, b), and that c is a point of (a, b). Then for each x in (a, b) there is a point $\bar{x}$ lying between x and c such that*

$$f(x) = P_n(x; c) + f^{(n+1)}(\bar{x})\frac{(x-c)^{n+1}}{(n+1)!}.$$

PROOF Fix a point $x = x_0$ in $[a, b]$. Suppose, for convenience that $x_0 < c$. On the interval $[x_0, c]$ define two new functions F and G as follows:

$$F(y) = f(x_0) - P_n(x_0; y),$$
$$G(y) = (x_0 - y)^{n+1}.$$

It is easily verified that F and G satisfy the hypotheses of Cauchy's theorem on $[x_0, c]$; hence there is a point $\bar{x}$ such that $x_0 < \bar{x} < c$ and

$$\frac{F(c) - F(x_0)}{G(c) - G(x_0)} = \frac{F'(\bar{x})}{G'(\bar{x})}.$$

Upon computing both sides of this equation, one finds that it reduces to

$$f(x_0) - P_n(x_0; c) = f^{(n+1)}(\bar{x})\frac{(x_0 - c)^{n+1}}{(n+1)!},$$

which is the required formula. ∎

Obviously if $c = 0$ the formulas become much simpler. When $c = 0$ the Taylor formula is called the *Maclaurin formula.*

If a function f has derivatives of all orders on some interval, then what we have proved is that we can approximate f with polynomials of arbitrarily high degree and at the same time have a formula for the amount of error in each approximation.

To illustrate this, consider the function $f(x) = \cos x$ defined, say, on $[-2\pi, 2\pi]$. This function has derivatives of all orders. The Taylor formula at $\pi/2$ with six terms and a remainder is easy to compute We get

$$f(x) = -\left(x - \frac{\pi}{2}\right) + \frac{\left(x - \frac{\pi}{2}\right)^3}{3!} - \frac{\left(x - \frac{\pi}{2}\right)^5}{5!} + R_5\left(x; \frac{\pi}{2}\right),$$

where $R_5(x, \pi/2) = (-\cos \bar{x})(x - \pi/2)^6/6!$ and $\bar{x}$ lies between x and $\pi/2$.

The corresponding Maclaurin formula is $f(x) = 1 - (x^2/2) + (x^4/4!) + R_5(x; 0)$, where $R_5(x; 0) = (-\cos \bar{x})(x^6/6!)$.

Notice that no matter what value $\bar{x}$ takes on, the absolute value $|R_5(x; 0)|$ of the error term is no greater than $|x|^6/6!$. Thus, for small values of x, $1 - (x^2/2!) + (x^4/4!)$ is a fairly good approximation to $\cos x$; the error in this approximation is, in absolute value, $\leq |2\pi|^6/6$. If one desires a closer approximation than this, evidently higher derivatives of $\cos x$ can be used and a better approximation is achieved.

EXERCISES

1. Obtain the first five terms in the Taylor polynomial of the following function and find also the remainder $R_5(x, c)$:

(a) $\sin x$, on $[-1, 1]$, $c = 0$
(b) $\tan x$, on $[-1, 1]$, $c = 0$
(c) $\ln (1 + x)$, on $[-\frac{1}{2}, \frac{1}{2}]$, $c = \frac{1}{4}$
(d) e^x, on $[1, 5]$, $c = 2$
(e) $\sinh x$, on $[-1, 1]$, $c = 0$

2. Determine the accuracy of the approximation

$$(1 + x^2)^{-1} \sim 1 - x^2 + x^4 - x^6 + x^8 \text{ in } [-\tfrac{1}{2}, \tfrac{3}{4}].$$

3. Determine the accuracy of the approximation

$$\sin x/x \sim 1 - (x^2/3!) + (x^4/5!) - (x^6/7!).$$

4. Determine $\int_0^1 \sin x/x$ with an error less than .05.

5. Determine arctan x with an error less than .05, given $0 < x < 1$. [*Hint:* arctan $x = \int_0^x (1/1 + t^2)\, dt$.]

4. DIFFERENTIATION AND INTEGRATION OF POWER SERIES

THEOREM 4.1 *Suppose the function f is defined by $f(x) = \sum a_n x^n$, where the power series has radius of convergence R. Then f is differentiable on the interval $(-R, R)$ and we have*

$$f'(x) = \sum_{n=1}^{\infty} n a_n x^{n-1}.$$

Remark. The theorem can be abbreviated to the statement that a power series may be differentiated *term by term* within the region of convergence.

Here is a sketch of the proof of the theorem. Let x_0 be the point of $(-R, R)$ at which we want to compute the derivative of f and let x be any point of $(-R, R)$.

Using Taylor's formula we have

$$x^n - x_0^n = (x - x_0)nx_0^{n-1} + \frac{(x - x_0)^2}{2} n(n-1)(\bar{x}_n)^{n-2},$$

where the point $\bar{x}_n$ (which depends on n) lies between x and x_0.

Thus the difference quotient of f at x_0 may be written

$$\frac{f(x) - f(x_0)}{x - x_0} = \frac{1}{x - x_0} \sum a_n(x^n - x_0^n)$$

$$= \frac{1}{x - x_0} \sum a_n[(x - x_0)nx_0^{n-1} + \frac{(x - x_0)^2}{2} n(n-1)(\bar{x}_n)^{n-2}]$$

$$= \sum na_n x_0^{n-1} + \sum \frac{n(n-1)}{2} a_n(x - x_0)(\bar{x}_n)^{n-2}.$$

The second term here may be written

$$(x - x_0) \sum \frac{n(n-1)}{2} a_n(\bar{x}_n)^{n-2}.$$

Notice here that when $n = 0$ or 1, the corresponding terms of the series have coefficient 0. Using the fact that $\sum a_n(\bar{x}_n)^{n-2}$ converges and that each $\bar{x}_n$ lies between x and x_0, it can be shown (see Exercise 1 of Section 2) that the series $\sum n(n-1)/2\, a_n(\bar{x}_n)^{n-2}$ converges. Its value depends only on the choice of the $\bar{x}_n$'s, which in turn depend on x; hence the value of this series is a function $S(x)$ of x. One then shows that as $x \to x_0$,

$$(x - x_0)S(x) \to 0.$$

Thus we have

$$\begin{aligned} f'(x_0) = \lim_{x \to x_0} \frac{f(x) - f(x_0)}{x - x_0} &= \sum n a_n x_0^{n-1} + \lim_{x \to x_0} (x - x_0)S(x) \\ &= \sum n a_n x_0^{n-1}. \quad \blacksquare \end{aligned}$$

The corresponding theorem for term-by-term integration can be stated as follows.

THEOREM 4.2 *Assume the hypotheses of Theorem* 4.1. *Then for every* x *in* $(-R, R)$ *we have*

$$\int_0^x f(t)\, dt = \sum \frac{a_n}{n+1} x^{n+1}.$$

PROOF. We use the fact that the two series $\sum a_n x^n$ and $\sum (a_n/n + 1)x^{n+1}$ have the same radius of convergence, R in this case. Let g be the function defined by the second of these power series. By the previous theorem $g'(x) = f(x)$ for every x in $(-R, R)$. Therefore, by the fundamental theorem of calculus,

$$\int_0^x f(t)\, dt = g(x) - g(0) = \sum \frac{a_n}{n+1} x^{n+1}. \quad \blacksquare$$

EXERCISES

1. Show that the function defined by the power series $\sum x^n/n!$ satisfies the differential equation $y' = y$. What is the function?
2. Compute the second derivative of the function defined by the power series $\sum (-1)^n(x^{2n+1}/2n + 1)$. Find a differential equation which this function satisfies and then find the function.

5. TAYLOR'S SERIES OF A FUNCTION

In Section 4 we showed that a function defined by a power series with positive radius of convergence is differentiable and by Exercise 1, Section 1, the derivative is defined by a power series with the same radius of convergence. The same result then applies to the derivative of the function and, proceeding inductively, we see that the original function has derivatives of all orders.

In this section we show that a partial converse is true: If a function has derivatives of all orders throughout an interval and if another condition, stated below, is satisfied, then the function is defined by a power series.

First, let us see what such a series must look like.

For generality we shall work with power series in $x - x_0$ rather than power series in x.

Suppose we have a series $\sum a_n(x - x_0)^n$ which defines a function f on the interval $(x_0 - R, x_0 + R)$, $R > 0$. Letting $x = x_0$, we see that $f(x_0) = a_0$. Differentiating once and setting $x = x_0$ we see that $f'(x_0) = a_1$. Repeating the process we find that $f''(x_0) = 2a_2$, $f'''(x_0) = 3{\cdot}2{\cdot}a_3$ and in general $f^{(n)}(x_0) = n!a_n$. Thus the values of the derivatives of f at x_0 completely determine the coefficients a_i, in the power series. This says there is only one candidate for the series we want: the one given in the following definition.

Suppose a function f has derivatives of all orders at a point x_0. The *Taylor's series for f* at x_0 is the series

$$\sum \frac{f^{(n)}(x_0)}{n!}(x - x_0)^n$$

(where $f^{(0)} = f$). In case $x_0 = 0$, the series is called the *Maclaurin series for f.*

What we proved in the preceding paragraph is this:

THEOREM 5.1 *If $\sum a_n(x - x_0)^n$ has positive radius of convergence and f is the function defined by this series on the interval of convergence, then this series is the Taylor's series of f.*

Before proceeding to the main theorem of this section, let us point out that the Taylor's series for a function f may not converge at every point where f is defined. This somewhat surprising fact is illustrated by the function $f(x) = 1/x$, which is defined for $x \neq 0$.

Let us compute the Taylor's series for f at $x_0 = 1$. It is not hard to see that the general formula for the derivative of f at x is

$$f^{(n)}(x) = (-1)^n n! x^{-(n+1)}.$$

Thus $f^{(n)}(1) = (-1)^n n!$ and the Taylor's series for f at $x_0 = 1$ is

$$\sum (-1)^n (x-1)^n.$$

Now this is just a geometric series and the interval of convergence is $(0, 2)$. Thus the series diverges for all $x \geq 2$ even though f is perfectly well defined for $x \geq 2$.

We remark that it can also happen that the Taylor's series for f converges at some value of x, but the sum of the series is not the value of f. Thus we are faced with the problem of finding conditions which tell us for what values of x the sum of the series at x exists and is equal to $f(x)$. The answer is given in the following theorem.

THEOREM 5.2 *Let f have derivatives of all orders in an interval (a, b) and let x_0 be a point of (a, b). The Taylor's series for f at x_0 converges at a point x_1 of (a, b), and its sum is $f(x_1)$ if and only if the sequence $\{R_n(x_1, x_0)\}$ converges to* 0.

Remark. Here $R_n(x_1, x_0)$ is the remainder term in the Taylor's formula for f at x_0 (see Section 3).

The proof of the theorem is easy. The nth partial sum of the Taylor's series for f at x_0 when evaluated at x_1 is just the nth-degree Taylor polynomial for f at x_0 evaluated at x_1, namely $P_n(x_1, x_0)$. By Taylor's formula we have, for each n,

$$f(x_1) - P_n(x_1, x_0) = R_n(x_1, x_0).$$

Thus the sequence $P_n(x_1, x_0)$ converges to $f(x_1)$ if and only if $R_n(x_1, x_0) \to 0$. ▮

Here are some examples, which illustrate the use of this theorem.

Example 1. The function $f(x) = \cos x$ has derivatives of all orders. Let us consider the Taylor's series for f at $x_0 = \pi/2$. The nth remainder term is given by

$$R_n\left(x, \frac{\pi}{2}\right) = \frac{f^{(n+1)}(\bar{x})}{(n+1)!}\left(x - \frac{\pi}{2}\right)^{n+1},$$

where $\bar{x}$ lies between x and $\pi/2$. Now for any $\bar{x}$ we have $|f^{(n+1)}(\bar{x})| \leq 1$ (why?). Hence for every x_1 and every n the absolute value of $R_n(x_1, \pi/2)$ is no greater than $|x_1 - (\pi/2)|^{n+1}/(n+1)!$. Let the constant $|x_1 - \pi/2|$ be denoted A. Thus we are concerned with the sequence $A^{n+1}/(n+1)!$. This converges to 0 for any value of A; hence the Taylor's series for $\cos x$ at $\pi/2$ converges and represents this function on the entire real line.

Example 2. The nth remainder term for the Taylor's series of $f(x) = 1/x$ is given by

$$R_n(x, 1) = \frac{f^{(n+1)}(1)}{(n+1)}(x-1)^{n+1} = (-1)^{n+1}(x-1)^{n+1}$$

For a fixed value of x, $\{R_n(x, 1)\}$ converges to 0 if and only if $|x - 1| < 1$. Thus this function is given by its Taylor's series at 1 only in the interval (0, 2).

EXERCISES

1. Find the Maclaurin series for each of the following functions and state the radius of convergence.

(a) e^{-ax} $(a > 0)$
(b) $\ln(1 + x)$
(c) $\sqrt{1 + \cos x}$ [*Hint:* $\cos x = 2\cos^2(x/2) - 1$.]
(d) $e^{\sin x}$

(e) $\ln \dfrac{1+x}{1-x}$

(f) $\tan^{-1}x$ (*Hint:* Note that $\tan^{-1}x = \int_0^x dt/1 + t^2$. Expand $1/1 + t^2$ in a Maclaurin series and integrate term by term.)

2. Find the Taylor's series for $\ln x$ about c and state the radius of convergence in terms of c.

3. Express the following as infinite series.

(a) $\displaystyle\int_0^x \frac{\ln(1+t)}{t}\,dt$ [See Exercise 1 (b).]

(b) $\displaystyle\int_0^x \frac{\tan^{-1} t}{t}\,dt$ [See Exercise 1 (f).]

APPENDIX I

Fields

1. INTRODUCTION

This appendix is designed to acquaint the reader with those algebraic notions which motivate the construction of the system of real numbers. The treatment will be limited to the presentation of definitions and examples, and the pace will be brisk.

2. BINARY OPERATIONS

A *binary operation* on a set X is a mapping from $X \times X$ into X. Some familiar examples of binary operations are addition, subtraction, and multiplication of real numbers and composition of functions from a set into itself.

It is customary to use symbols such as $+$, $-$, $\cdot$, and $\circ$ for binary operations, and if $*$ is a binary operation on X, then the image of the pair (x, y) under $*$ is denoted $x * y$ rather than $*((x, y))$.

We observe that division, $\div$, is a binary operation on the set of nonzero rational numbers or the set of nonzero real numbers but not on the set of all rational or real numbers (for example, $2 \div 0$ is not defined). Similarly, addition is not a binary operation on the set of irrational numbers since the sum of two irrationals may be rational.

A binary operation $*$ on X is called *commutative* if $x * y = y * x$ for all x and y in X; it is called *associative* if $x * (y * z) = (x * y) * z$ for all x, y, and z. If we think of a binary operation on X as a method of combining ordered pairs of elements of X to get new elements, then to say that the operation is commutative is to say that the order of combination is irrelevant. The operation is associative if it does not matter how we group the elements (as long as the order in which they appear is kept fixed).

Addition and multiplication are both commutative and associative. Subtraction is neither commutative nor associative, as the reader can easily check. If $\mathscr{F}$ is the collection of all functions from a set having at least two elements into itself, then composition of functions in $\mathscr{F}$ is associative but not commutative. Commutative but not associative operations are generally somewhat artificial in nature; an example is the operation $*$ defined on the set of real numbers by $x * y =$ the largest integer less than or equal to $x + y$. For this operation we have $(\frac{1}{3} * \frac{1}{3}) * \frac{2}{3} = 0$; but $\frac{1}{3} * (\frac{1}{3} * \frac{2}{3}) = 1$.

Let $*$ be a binary operation on a set X. What meaning can we attach to an expression such as $x_1 * x_2 * \cdots * x_n$, where the x_i are in X? In general one can compute a value for this expression by introducing parenthesis, that is, by grouping the terms. For $n = 3$ there are two such groupings: $(x_1 * x_2) * x_3$ and $x_1 * (x_2 * x_3)$. For $n = 4$ there are five possible groupings, and so on. As shown above, even for $n = 3$, different groupings may lead to different answers. However, if $*$ is associative, then all groupings yield the same answer. For $n = 3$ this is merely the definition of associativity. For larger values of n, the corresponding statement is cumbersome to formulate and to prove. A rough statement goes as follows:

GENERALIZED ASSOCIATIVITY LAW. *Let $*$ be an associative binary operation on X and let $x_1, x_2, \ldots, x_n$ be elements of X. Let y and z be elements of X obtained by grouping terms in the expression $x_1 * \cdots * x_n$ in two possibly different ways; then $y = z$.*

3. IDENTITY ELEMENTS AND INVERSES

Throughout this section, let $*$ be an associative binary operation on a set X. We say that an element e of X is a $*$ *identity* provided that $e * x = x * e = e$ for every x in X.

For example, letting $+$ and $\cdot$ be the usual operations on the set of real numbers, we observe that 0 and 1 are, respectively, the $+$ and $\cdot$ identities. If $+$ is considered as a binary operation on the set of positive real numbers, there is no $+$ identity.

It is easy to see that *if there is a $*$ identity in X, then it is unique*. For suppose e and f satisfy the definition of a $*$ identity; then $e = e * f = f$ (justify the equality signs).

Suppose now that there is a $*$ identity e in X. Let x be an element of X; an element y of X is called a $*$ *inverse* of x in case $x * y = y * x = e$. If y and z are $*$ inverses for x, then $y = y * e = y * (x * z) = (y * x) * z = e * z = z$. This shows that if x has a $*$ inverse, then it has only one; that is, $*$ *inverses are unique when they exist.*

In the case of the operation $+$ on the set of all real numbers, the $+$ inverse of a number x is simply $-x$. However, if $+$ is considered as a binary operation on the set of nonnegative reals, then although there is a $+$ identity, 0, very few elements (in fact, only one) have $+$ inverses.

The reader might wish to investigate the following examples for the existence of identities and inverses.

Example A. Multiplication on the set of all reals, on the set of rational numbers and on the set of integers.

Example B. Composition on the set of all functions from some fixed set into itself and on the set of all 1-1 functions from some set onto itself.

Example C. The binary operations $\oplus$ and $\odot$ defined on the set of real numbers by

$x \oplus y = x + y - 1$ and $x \odot y = x + y - x \cdot y$, where $+$, $-$, and $\cdot$ have their usual meanings.

Example D. The operations $\cup$ and $\cap$ defined on the collection of *all* subsets of a fixed set.

4. FIELDS

A *field* is an ordered triple $(X, +, \cdot)$ where X is a set having at least two elements and $+$ and $\cdot$ are commutative, associative binary operations on X satisfying the following requirements:

4.1. There is a $+$ identity, usually denoted 0 and every element x of X has a $+$ inverse which is usually denoted $-x$.

4.2. There is a $\cdot$ identity, usually denoted 1, and every element x of X, except 0, has a $\cdot$ inverse which is usually denoted x^{-1}.

4.3. For every x, y, and z in X we have $x \cdot (y + z) = x \cdot y + x \cdot z$. This is called the *distributive law.*

We emphasize that $+$ and $\cdot$ are not necessarily ordinary addition and multiplication of numbers; indeed, X is merely an abstract set of objects. The use of the symbols $+$ and $\cdot$ is motivated by several considerations. They are easy to write and read and they behave much as addition and multiplication of real numbers; this last point can be illustrated by some of the computational rules which follow directly from the axioms for a field.

Before stating these, let us point out that if x and y are elements of a field, then the sum of x and the $+$ inverse of y is usually written $x - y$ rather than $x + (-y)$.

Let $(X, +, \cdot)$ be any field and let x, y, z, and w be elements of X. Then:

1. $x \cdot 0 = 0$.
2. $-(-x) = x$.
3. $-(x + y) = -x - y$.
4. $x \cdot (-y) = (-x) \cdot y$.
5. $(-1) \cdot x = -x$.
6. $-(x^{-1}) = (-x)^{-1}$ if $x \neq 0$.
7. $(x^{-1})^{-1} = x$ if $x \neq 0$.
8. $(x \cdot y)^{-1} = x^{-1} \cdot y^{-1}$.
9. $x \cdot y^{-1} + z \cdot w^{-1} = (x \cdot w + y \cdot z)(y \cdot w)^{-1}$ if $y \neq 0 \neq w$.

We shall prove 1 and 8 to illustrate how to use the axioms.

To prove 1 we observe that $x \cdot 0 = x \cdot (0 + 0) = x \cdot 0 + x \cdot 0$. The first equality follows from the fact that $0 + 0 = 0$ and the second from the distributive law. Letting $y = x \cdot 0$ we thus have shown that $y = y + y$. Adding $-y$ to both sides we get. $y + (-y) = y + y + (-y)$ or $0 = y$, as was to be shown.

To prove 8 we use commutativity and associativity of $\cdot$ as follows:

$$\begin{aligned}(x \cdot y) \cdot (x^{-1} \cdot y^{-1}) &= (x \cdot (y \cdot x^{-1})) \cdot y^{-1} \\ &= ((x \cdot x^{-1}) \cdot y) \cdot y^{-1} \\ &= (x \cdot x^{-1}) \cdot (y \cdot y^{-1}) \\ &= 1 \cdot 1 = 1.\end{aligned}$$

This shows that $x^{-1} \cdot y^{-1}$ is the $\cdot$ inverse of $x \cdot y$; in symbols, $(x \cdot y)^{-1} = x^{-1} \cdot y^{-1}$.

We urge the reader to try proving the remaining results.

The requirement that the field have at least two elements is mainly for nontriviality. Among other things it implies that in any field, $1 \neq 0$. For if $1 = 0$, then for every x in the field we have $x = x \cdot 1 = x \cdot 0 = 0$. Thus if 1 were 0, then the field would have only one element, contradicting the definition.

Here are some examples of fields.

Example A. The set of rational numbers with the usual operations.

Example B. The set of real numbers with the usual operations

Example C. The set of real numbers with the operations $\oplus$ and $\odot$ defined as in Example C of Section 3.

Example D. The set C of all ordered pairs of real numbers with the operations $\oplus$ and $\odot$ defined as follows:

$$(x, y) \oplus (z, w) = (x + z, y + w),$$
$$(x, y) \odot (z, w) = (x \cdot z - y \cdot w, y \cdot z + x \cdot w),$$

where the operations on the right are the usual ones for real numbers.

Assuming both $\oplus$ and $\odot$ are associative, the reader should verify the other axioms for a field in this example. The symbols $\hat{0}$ and $\hat{1}$ are used to denote the $\oplus$ and $\odot$ identities in Example C to distinguish them from $+$ and $\cdot$ identities in the set of real numbers. The field $(C, \oplus, \odot)$ has the amusing property that some elements appear to have "negative" squares. For example, if $p = (0, 1) \in C$, then $p \odot p = -\hat{1}$.

5. ORDERED FIELDS

Let $(X, +, \cdot)$ be a field. A subset P of X is called a set of *positive elements* for X provided:

P1. P is closed under addition and multiplication; that is, if x and y are in P, then so are $x + y$ and $x \cdot y$.

P2. For every x in X exactly one of the following holds:

$$x \in P, \qquad x = 0, \qquad \text{or} \qquad -x \in P.$$

If the field $(X, +, \cdot)$ has a set P of positive elements, then the "square," $x \cdot x$ of every nonzero element x, belongs to P. To see this, note that either $x \in P$ or else $-x \in P$; hence $x \cdot x \in P$ or

$(-x) \cdot (-x) \in P$. But using facts 2) and 4) of Section 4 we see that $x \cdot x = (-x) \cdot (-x)$, hence $x \cdot x \in P$. In particular, 1 is always in P.

It can also be shown that if x is in P, then so is x^{-1}.

Of the examples listed in the preceding section, A and B have sets of positive elements (the usual ones). Example C has a set of positive elements: all real numbers which are strictly less than 1 in the usual ordering of the reals. Example D cannot have a set of positive elements. For, the field Q contains an element p such that $p \odot p = -\hat{1}$. Thus if Q had a set P of positive elements, we would have $-\hat{1}$ and $\hat{1}$ in P, contradicting condition P2.

Suppose now that the field $(X, +, \cdot)$ has a set P of positive elements. We define a relation $<$ in X by the conditon $x < y$ provided $y - x \in P$. It is an easy matter to check that $<$ is a linear ordering of X. (See Chapter 3 for the definition and basic properties of linear orderings.) Also this ordering behaves nicely with respect to the field operations. We can summarize the essential properties of $<$ as follows:

O1. For every x in X either $x > 0$, $x = 0$ or $x < 0$.
O2. If $x > 0$ and $y > 0$, then $x + y > 0$.
O3. If $x > 0$ and $y > 0$, then $x \cdot y > 0$.

From these properties follow a host of useful rules concerning inequalities. For example, if $x < y$ and z is arbitrary, then $x + z < y + z$, and if $x < y$ and $z > 0$, then $x \cdot z < y \cdot z$.

It turns out that if the field X admits a partial ordering which satisfies the rules O1, O2, and O3, then the set $P = \{x \in X \mid 0 < x\}$ is a set of positive elements for X; that is, P will satisfy axioms P1 and P2. Thus the existence of a set of positive elements is equivalent to the existence of a partial ordering satisfying the three axioms given above.

A field which admits a set of positive elements (and hence an ordering like the above) is called an *ordered field*. The field of reals, the field of rationals, and the funny field of Example **C** above are ordered fields, while the field of Example **D** is not.

Now suppose $(X, +, \cdot)$ is an ordered field and let $<$ be an ordering satisfying the above axioms. We say that X is *Archimedean* provided that if x and y are elements of X and $x > 0$, then there is a positive integer N such that Nx (which is defined to be $x+x+\cdots+x$ with N summands) is greater than y. All the ordered fields given above are Archimedean. There do, however, exist non-Archimedean ordered fields.

We close with one final notion. It is this idea which enables us to distinguish the field of rationals from the field of real numbers. Let $(X, +, \cdot)$ be an ordered field with ordering $<$ as above. A subset A of X will be called a *Dedekind cut*, or just a *cut*, provided that:

1. $A \neq \varnothing$ and $A \neq X$.
2. If x is in A and $y<x$, then $y \in A$.
3. A has no largest element; that is, if x is in A, then there exists y in A such that $x<y$.

The field is said to have the Dedekind property or to be *Dedekind complete* provided that if A is any cut in X, then there is an element a in X (necessarily unique) such that

$$A=\{x \in X \mid x<a\}.$$

The field of rational numbers does not have this completeness property. For example, the set $A=\{x \mid x \leq 0\} \cup \{x \mid x^2<2\}$ is a cut, but there is no rational number a such that $A=\{x \mid x<a\}$. This can be proved using the fact that $\sqrt{2}$ is not a rational number.

On the other hand, as we show in Appendix 2 the field of all real numbers is Dedekind complete. Indeed as we pointed out in Chapter 5, it is the need in analysis for completeness (in this technical sense) which motivates the very construction of the field of real numbers which is complete as an extension of the field of rationals.

APPENDIX 2

Construction of the Real Numbers by Dedekind Cuts

1. INTRODUCTION

In Chapter 5 we gave a quick outline of the basic properties of the real numbers and a hint of how one might construct the set of real numbers beginning with the set of rationals.

This appendix is devoted to a rigorous construction of the ordered field of real numbers assuming the existence of the ordered field of rationals.

To begin with, then, we have the ordered field Q of rational numbers. Formally Q may be defined as the collection of equivalence classes of pairs (p, q), where p and q are integers with $q \neq 0$, and the equivalence relation is defined by $(p, q) \approx (r, s)$ if and only if $sp = qr$ (ordinary multiplication of integers). Thus the pairs $(2, 3)$, $(4, 6)$, and $(-10, -15)$ represent the same rational number. Of course, it is customary to write p/q rather than (p, q).

One has the usual rules for addition and multiplication of rational numbers and the usual ordering. To summarize, the rational numbers with the usual operations and ordering form an Archimedean ordered field (see Appendix 1 for precise definitions). However, Q is not Dedekind complete.

What we shall construct is an Archimedean ordered field which is complete. This field will be called the field of real numbers.

The initial steps of the construction will be presented in some detail. These include the definition of those objects (cuts) which eventually emerge, as real numbers, the order relation between these cuts, and the definition of addition. As more structure is added and more complicated relations arise, we decrease the amount of technical detail and concentrate on the highlights of the process.

2. DEDEKIND CUTS

Let Q denote the field of rational numbers with the usual ordering. A subset A of Q is called a *Dedekind cut* or simply a *cut*, provided that it has the following properties:

1. $A \neq \varnothing$ and $A \neq Q$.

2. If $p \in A$ and $q < p$, then $q \in A$.
3. A has no largest element; that is, if $p \in A$, then there exists $q \in A$ such that $p < q$.

Here are two examples of cuts.

Example A. Let r be a fixed rational number and let $A_r = \{p \in Q \mid p < r\}$. Clearly A_r has properties 1 and 2. If $p \in A_r$, then the rational number $q = (p + r)/2$ satisfies $p < q < r$; hence A_r also has property 3.

The cut A_r will be called the cut *determined by* the rational number r and, for brevity, this cut will be denoted $\hat{r}$. Thus, for example, $\hat{0}$ is the set of negative rationals.

Example B. Let $A = \{p \in Q \mid p^2 < 2\} \cup \{\text{negative rationals}\}$. Since $0 \in A$ and $2 \in A$, property 1 holds. Suppose $0 < q < p$, where $p \in A$; since $q^2 < p^2 < 2$ we have $q \in A$, which shows that property 2 holds. To verify the third property suppose $p^2 < 2$. Choose a rational number r such that $0 < r < 1$ and $r < (2-p^2)/2p + 1$. Put $q = p + r$; then $q > p$ and $q^2 < 2$.

Let R denote the collection of all cuts. For reasons which will become clear later, we will denote set inclusion by $<$ rather than by $\subset$; thus $<$ is a partial ordering on R and we have the associated orderings $\leq$, $>$, and $\geq$.

Presently we are going to define algebraic operations in R and investigate the relationship between these and the ordering $<$. It will be useful to have the following alternative characterization of the order relation:

THEOREM 2.1 *If A and B are cuts, then $A < B$ if and only if $B - A \neq \varnothing$.*

The proof in one direction is trivial since $A < B$ means $A \subset B$. For the converse, suppose there is a rational $p \in B - A$. Since p does not belong to A we must have $q < p$ for every q in A (why?), and hence, by condition (2), every q in A belongs to B. This shows that $A < B$. ▮

In general, of course, set containment yields only a partial ordering. However, in our case something stronger holds.

THEOREM 2.2 *The relation $<$ is a linear ordering (defined in Exercise 6, Section 6, Chapter 3) of R.*

Verification of this theorem is simple but instructive and we leave it as an exercise.

3. ADDITION OF CUTS

If A and B are cuts, then $A + B$, read "A plus B," is defined to be the set of all rational numbers of the form $p + q$, where $p \in A$, $q \in B$;

$$A + B = \{p + q \mid p \in A, q \in B\}.$$

We first note that the set $A + B$ is itself a cut. It is certainly not empty since neither of A and B is empty. To see that $A + B \neq Q$, note that since $A \neq Q$ and $B \neq Q$ there exist rationals x and y such that $p < x$ for every p in A and $q < y$ for every q in B. It follows that if $r = p + q$ is any element of $A + B$, then $r < x + y$. Thus $x + y \notin A + B$ and $A + B$ has property (1).

Property 2 follows easily from the corresponding properties for A and B. To prove that $A + B$ has property 3 requires some ingenuity; the reader will enjoy trying this for himself.

What these remarks show is the $+$ is a binary operation on R. More is true, namely:

THEOREM 3.1. *The operation $+$ is associative and commutative, there is a $+$ identity and every element of R has a $+$ inverse.*

Associativity and commutativity follow quickly from the fact that addition in Q is associative and commutative. There is an obvious candidate for identity, $\hat{0}$, the cut determined by 0 in Q. To verify that $\hat{0}$ is the identity, notice that if A is a cut and $p \in A$ and $q \in \hat{0}$, then $p + q < p$, hence $p + q \in A$. Thus $A + \hat{0} \leq A$. On the other hand, if $p \in A$, choose $q \in A$ such that $p < q$. Then the rational number $r = p - q$ is in $\hat{0}$ and $q + r = p$ is in $A + \hat{0}$. So $A \leq A + \hat{0}$, and by the previous inequality we must have $A = A + \hat{0}$, which shows that $\hat{0}$ is the identity in $(R, +)$.

To prove the existence of inverses in $(R, +)$ we will need the following.

LEMMA 3.2. *Let A be a cut and let r be a positive rational. Then there exist p in A and q in $Q - A$ such that $q - p = r$ and q is not the smallest element of $Q - A$.*

PROOF. Let p_0 be any element of A and let $p_n = p_0 + nr$ for $n = 0, 1, 2, \ldots$. There exists a unique integer $N \geq 0$ such that $p_n \in A$ for $n \leq N$ and $p_n \in Q - A$ for $n > N$. If p_{N+1} is the smallest element of $Q - A$, then the rationals $p = p_N + (r/2)$ and $q = p_{N+1} + (r/2)$ satisfy the conclusion of the lemma. If p_{N+1} is not the smallest element of $Q - A$, then we take $p = p_N$, $q = p_{N+1}$. (Pictures are quite useful in understanding this proof.) ∎

Returning to the theorem, let A be a cut and let $B = \{q \in Q \mid -q \in Q - A$ and $-q$ is not the smallest element in $Q - A\}$. B is not empty since $Q - A$ has at least two elements; also $B \neq Q$, since, if so, every element of Q would belong to $Q - A$, contradicting the fact that A is a cut. If $p \in B$ and $q < p$, then $-q > -p$, from which it follows that $q \in B$. Finally, suppose $p \in B$; then there exists an element q_0 of $Q - A$ with $q_0 < -p$. Let $q = (p - q_0)/2$; then $q_0 < -q < -p$, which implies that $q \in B$ and $p < q$. Thus we have verified that B is a cut.

We now show that $A + B = \hat{0}$. It is easy to see that $A + B \leq \hat{0}$, for if $p \in A$ and $q \in B$, then $-q \notin A$, which implies that $p < -q$ and hence $p + q < 0$. Conversely, suppose r is an element of $\hat{0}$. By the lemma proved above, there exist $p \in A$ and $q \in B$ such that $-q - p = -r$, whence $r = p + q \in A + B$. So $\hat{0} \leq A + B$, which with the inequality proved above shows that $A + B = \hat{0}$.

Thus B is indeed the inverse of A in $(R, +)$ and the proof is complete. ∎

We proved Theorem 3.1 in great detail to illustrate the techniques one uses in constructing the real number system by Dedekind cuts. For the remainder of this appendix we shall omit most of the proofs, restricting our attention to the construction process itself.

4. MULTIPLICATION OF CUTS

Since eventually we wish to make R into a field we must define multiplication in R. This is done in two steps.

To begin with, suppose A and B are cuts and $A \geq \hat{0}$, $B \geq \hat{0}$. We define $A \cdot B$ to be the set $\{pq \mid p \geq 0, q \geq 0, p \in A, q \in B\}$. It is not hard to verify that $A \cdot B$ is a cut.

The *absolute value* $|A|$ of a cut A is defined to be A if $A \geq \hat{0}$ and $-A$ if $A < \hat{0}$ (here $-A$ means the additive inverse of A in R). Thus $|A|$ is a cut.

The product $A \cdot B$ of two arbitrary cuts is then defined as follows:

$$A \cdot B = \begin{cases} |A| \cdot |B| & \text{if } A \geq \hat{0} \text{ and } B \geq \hat{0} \text{ or if } A \leq \hat{0} \text{ and } \\ & B \leq \hat{0}, \\ -(|A| \cdot |B|) & \text{otherwise.} \end{cases}$$

The product is unambiguously defined and it is certainly a cut. It is a tedious and at times tricky matter to show that $(R, +, \cdot)$ is a field. After showing that R is commutative and associative and that the distributive laws hold, it remains to show that there is a multiplicative identity and that each cut except $\hat{0}$ has a multiplicative inverse. It is not hard to see that $\hat{1}$, the cut determined by the rational number 1, is the multiplicative identity. If A is a cut and $A > \hat{0}$, then we define B to be the union of all cuts of the form $\hat{p}$, where $0 < 1/p$ and $1/p \in A$. It is easy to see that B is a cut (the union of cuts is a cut if it is not all of Q) and with a little more work one can show that $A \cdot B = \hat{1}$. Thus every cut A, with $A > \hat{0}$, has a multiplicative inverse which is denoted A^{-1}. Given an arbitrary nonzero cut, A, let B be the cut given by

$$B = \begin{cases} A^{-1} & \text{if } A > \hat{0}, \\ -|A|^{-1} & \text{if } A < \hat{0}. \end{cases}$$

It turns out that $A \cdot B = \hat{1}$; hence every nonzero cut A has an inverse and $(R, +, \cdot)$ is a field, called *the field of real numbers*.

There is a natural mapping from Q into R, denoted by $f(p) = \hat{p}$. It is not hard to see that f is 1-1, "preserves" the algebraic operations, and preserves order. More precisely, if p and q are rationals, then $f(p+q) = f(p) + f(q)$, $f(p \cdot q) = f(p) \cdot f(q)$ and $p < q$ if and only if $f(p) < f(q)$. Thus, identifying each rational with the image under f we may think of the rationals as a "subfield" of the field R.

5. ORDER PROPERTIES OF THE REAL NUMBERS

Let R_p denote the set of elements A in R such that A contains at least one positive rational. It is easy to verify that R_p is a set of positive elements for R (see Section 5, Appendix 1). Hence R_p induces a partial ordering $\prec$ on R, $A \prec B$, provided $B - A \in R_p$ (here the minus sign means the algebraic difference of B and A in the field R). It turns out that $\prec$ agrees with the linear ordering $<$ of R defined in Section 2; that is, $A \prec B$ if and only if $A \prec B$. Thus, *R with the natural ordering $<$ is an ordered field.*

The "usual" relations between the ordering of real numbers and the algebraic operations now follow from the general theory as outlined in Appendix 1.

Observe that by construction *the rational cuts are dense in R*; more precisely, if A and B are elements of R with $A < B$, then there is a rational number p such that $A < \hat{p} < B$. To see this, note that if $A \prec B$, then there is a rational number q belonging to $B - A$. Since q is not the largest element of B, there is a rational number p in B such that $q < p$. We then have $A < \hat{p}$ and $\hat{p} < B$ (Why is it not necessarily true that $A < \hat{q} < B$?)

We can also show that R is *Archimedean*; that is, given elements A and B of R with $A > \hat{0}$, there exists an integer n such that $nA > B$. It is by no means difficult to prove that the ordered field Q of rational numbers is Archimedean. Given elements A, B of R as above, choose a rational p in A such that $0 < p$ and choose a rational q not in B. Since Q is Archimedean there is an integer n such that $np > q$. Recall that the mapping f which takes p to $\hat{p}$ is an order-preserving function. Hence we have $\hat{q} < n\hat{p}$. Now p was chosen so that $\hat{p} < A$, hence by the usual properties of addition in an ordered field $n\hat{p} < nA$. On the other hand, q was chosen so that $B < \hat{q}$. Putting these inequalities together, we see that $B < nA$, as desired.

Finally we show that the field we have constructed is *Dedekind complete*. Let $\mathscr{A}$ be a cut in R; thus $\mathscr{A}$ is a collection of elements of R (which are themselves cuts in Q) satisfying

1. $\mathscr{A} \neq \varnothing$ and $\mathscr{A} \neq R$.
2. If $X \in \mathscr{A}$ and $Y < X$, then $Y \in \mathscr{A}$.
3. $\mathscr{A}$ has no largest element.

We now consider $\mathscr{A}$ as a collection of subsets of D and let A be the union of $\mathscr{A}$. Thus A is a set of rational numbers. It is easy to verify that A is a cut in the set of rationals or, in other words, A is an element of R. Moreover, one easily shows that this element of R determines the cut $\mathscr{A}$ in the sense that $\mathscr{A} = \{X \in R \mid X < A\}$. Thus R is Dedekind complete.

It can be shown that as mathematical objects any two complete Archimedean ordered fields are essentially the same; that is, there is a 1-1 map which preserves the binary operations and the order structure from field onto the other. Any such field is called the field of real numbers.

APPENDIX 3

Construction of the Real Numbers by Cauchy Sequences

1. INTRODUCTION

In Appendix 2 we outlined a construction of the real numbers using Dedekind Cuts. Here we present a different approach to the same problem.

The basic idea can be described as follows. Every Cauchy sequence of rational numbers converges to a real (but not necessarily rational) number and, conversely, every real number can be written as the limit of a (necessarily Cauchy) sequence of rational numbers. Thus there is a natural correspondence between the set of all Cauchy sequences of rationals and the set of all real numbers. This observation is the basis for the construction process outlined in this appendix.

2. THE CONSTRUCTION

Let Q be the set of rational numbers with the usual algebraic and order properties. Recall that a sequence $\{x_n\}$ in Q is Cauchy provided that for any positive rational ε there is an integer N such that $|x_n - x_m| < \varepsilon$ if $n,m \geq N$. For example, any constant sequence is Cauchy and if $\{x_n\}$ is a sequence in Q converging to x in Q then $\{x_n\}$ is Cauchy.

Letting $\mathscr{C}$ denote the collection of all Cauchy sequences in Q, we observe that $\mathscr{C}$ is closed under termwise addition and multiplication. More precisely suppose $\{a_n\}$ and $\{b_n\}$ are in $\mathscr{C}$, then so are their sum $\{a_n\} + \{b_n\} = \{a_n + b_n\}$ and their product $\{a_n\} \cdot \{b_n\} = \{a_n b_n\}$. In particular if $\{a_n\}$ is in $\mathscr{C}$ and r is a rational number (which can be thought of as a constant sequence) then the sequence $r \cdot \{a_n\} = \{ra_n\}$ is again in $\mathscr{C}$.

The operations of addition and multiplication in $\mathscr{C}$ are associative and indeed the triple $(\mathscr{C}\ +, \cdot)$ satisfies all of the axioms for a field except that nonzero elements need not have multiplicative inverses. To remedy this we introduce an equivalence relation in $\mathscr{C}$. Let $\sim$ be the relation defined on $\mathscr{C}$ by $\{a_n\} \sim \{b_n\}$ provided $\{a_n - b_n\}$ converges to zero.

We point out that if $\{a_n\} \sim \{b_n\}$ and $a_n \to x$ in Q then, also, $b_n \to x$ in Q. A very important fact and one which we shall use several times in the construction is that if $\{a_n\}$ is in $\mathscr{C}$ and $\{b_n\}$ is a subsequence of $\{a_n\}$, then $\{b_n\}$ is in $\mathscr{C}$ and $\{b_n\} \sim \{a_n\}$. This follows from the order preserving properties of selection functions.

THEOREM 2.1. $\sim$ *is an equivalence relation on* $\mathscr{C}$.

The proof of transitivity is an application of the triangle inequality. Reflexivity and symmetry are trivial.

We are now ready to define the set which will eventually emerge as the set of real numbers. The collection of all equivalence classes of elements of $\mathscr{C}$ (induced by $\sim$) will be denoted R. We use small greek letters $\alpha, \beta, \gamma, \ldots$, to denote equivalence classes. By the remark preceeding Theorem 2, if α is in R then either all the sequences in α converge to some element of Q or none of them does.

The addition and multiplication in $\mathscr{C}$ induce corresponding operations in R. Before giving precise definitions we need a lemma.

LEMMA 2.2. *Suppose* $\{a_n\} \sim \{c_n\}$ *and* $\{b_n\} \sim \{d_n\}$ *in* $\mathscr{C}$. *Then* $\{a_n + b_n\} \sim \{c_n + d_n\}$.

PROOF. We want to show that the sequence $\{(a_n+b_n)-(c_n+d_n)\}$ converges to zero. This sequence can be written as $\{a_n - c_n\} + \{b_n - d_n\}$. Each of these sequences converges to zero, hence, so does their sum. ∎

Given elements α and β of R define $\alpha + \beta$ as follows: choose any element $\{a_n\}$ in α and any element $\{b_n\}$ in β. Then $\alpha + \beta$ is the equivalence class containing $\{a_n + b_n\}$. Lemma 2.2 guarantees that $\alpha + \beta$ is well defined; that is, the equivalence class $\alpha + \beta$ does not depend on which sequences in α and β we choose in order to compute an element of $\alpha + \beta$.

It is an easy matter to check that $+$, called addition, is a commutative, associative operation in R; that θ, the equivalence class containing the constant zero sequence, is the additive identity and that every α in R has an additive inverse. The reader should verify these facts as exercises.

Multiplication in R is defined in the following way. If α and β are in R then $\alpha \cdot \beta$ is the equivalence class containing the sequence $\{a_n b_n\}$ where $\{a_n\}$ is any element of α and $\{b_n\}$ is any element of β. That multiplication is well defined follows from the next result.

LEMMA 2.3. *If* $\{a_n\} \sim \{c_n\}$ *and* $\{b_n\} \sim \{d_n\}$, *then* $\{a_n b_n\} \sim \{c_n d_n\}$.

PROOF. Since $\{a_n\}$ and $\{d_n\}$ are Cauchy, they are bounded (see the proof of Theorem 6.1, Chapter 5); let M be a rational number such that $|a_n| \leq M$ and $|d_n| \leq M$ for all n.

Given a positive rational ε choose an integer N such that if $n \geq N$ then $|a_n - c_n| < \varepsilon/2M$ and $|b_n - d_n| < \varepsilon/2M$. Such an integer exists because the sequence $\{a_n - c_n\}$ and $\{b_n - d_n\}$ converge to zero.

Now, if $n \geq N$, then

$$\begin{aligned} |a_n b_n - c_n d_n| &= |a_n(b_n - d_n) + d_n(a_n - c_n)| \\ &\leq |a_n|\,|b_n - d_n| + |d_n|\,|a_n - c_n| \\ &< M.\,(\varepsilon/2M) + \cdot\,(\varepsilon/2M) = \varepsilon. \end{aligned}$$

Thus $\{a_n b_n - c_n d_n\}$ converges to zero, that is, $\{a_n b_n\} \sim \{c_n d_n\}$. ∎

Again we leave to the reader, as an exercise, the task of verifying that multiplication in R is commutative and associative and that the equivalence class φ containing the constant sequence $\{1\}$ is the multiplicative identity. The distributive law also holds in R, that is, if α, β and γ are in R then $\alpha \cdot (\beta + \gamma) = \alpha \cdot \beta + \alpha \cdot \gamma$. Thus to see that R is a field it remains only to show that every element except θ, the additive identity, has a multiplicative inverse.

Let α be a member of $R - \{\theta\}$ and let $\{a_n\}$ be any element of α. We assert that the terms of $\{a_n\}$ are bounded away from zero for sufficiently large n. That is, we assert that there is a positive rational δ and an integer N such that if $n \geq N$, then $|a_n| \geq \delta$. If this were false then one could extract a subsequence of $\{a_n\}$ converging to zero (the reader should do this). But then since $\{a_n\}$ is Cauchy it would follow that $\{a_n\}$ itself converges to zero contradicting the hypothesis that $\{a_n\}$ is not an element of θ.

The construction of a multiplicative inverse for $\{a_n\}$ now proceeds as follows. Let δ and N be the numbers obtained in the preceding paragraph and let $\{b_n\}$ be the subsequence given by $b_n = a_{n+N}$. Thus $b_1 = a_{N+1}$, $b_2 = a_{N+2}$ and so on; note that $|b_n| > \delta$ for all n and that, as noted earlier, $\{a_n\} \sim \{b_n\}$. Let $\{c_n\}$ be the sequence given by $c_n = 1/b_n$.

LEMMA 2.4. *The sequence $\{c_n\}$ constructed above is Cauchy and if γ is the equivalence class containing $\{c_n\}$ then $\alpha \cdot \gamma = \varphi$.*

PROOF. Given a positive number ε, choose N_1 so large that if $n, m \geq N_1$ then $|b_n - b_m| < \varepsilon\delta^2$. Let N_2 be the larger of N_1 and N. Then if $n,m \geq N_2$ we have $|c_n - c_m| = |1/b_n - 1/b_m| = |(b_n - b_m)/b_n b_m| < \varepsilon\delta^2/\delta^2 = \varepsilon$. This shows that $\{c_n\}$ is Cauchy. ▮

To prove that $\gamma \cdot \alpha = \varphi$, we merely note that $\gamma \cdot \alpha$ contains $\{c_n\} \cdot \{b_n\} = \{c_n b_n\} = \{1\}$.

This lemma completes the proof that $(R, +, \cdot)$ is a field. Henceforth the elements of R (the equivalence classes of Cauchy sequences of rationals) will be called *real numbers*. The additive inverse of an element α of R is denoted $-\alpha$ and its multiplicative inverse is denoted α^{-1}.

We end the detailed part of this appendix by indicating how one defines the set of positive elements for R.

Let us call a sequence $\{a_n\}$ in $\mathscr{C}$ *positive* if there is a subsequence $\{b_n\}$ of $\{a_n\}$ and an ε such that $b_n \geq \varepsilon > 0$ for all n; and *negative* if there is a subsequence $\{b_n\}$ of $\{a_n\}$ such that $b_n \leq \varepsilon < 0$ for all n. With some work, it can be shown that if $\{a_n\}$ is in $\mathscr{C}$ then exactly one of the following holds:

1. $\{a_n\}$ converges to zero.
2. $\{a_n\}$ is positive.
3. $\{a_n\}$ is negative (in which case $\{-a_n\}$ is positive).

Moreover if $\{a_n\} \sim \{b_n\}$ then $\{a_n\}$ satisfies 1. or 2. or 3. if and only if $\{b_n\}$ does.

THEOREM 2.5 *Let R_p be the set of elements in R which contain a positive sequence, then R_p is a set of positive elements for the field $(R, +, \cdot)$.*

PROOF. By the last remark made above whether or not an element α of R belongs to R_p is determined by looking at any sequence in α; in other words, membership in R_p is well defined.

Next we observe that the other remarks made above imply that if α is in R then exactly one of the following holds:

1. $\alpha = \theta$.

2. α is in R_p.
3. $-\alpha$ is in R_p.

It is easy to verify that the (termwise) sum and product of positive sequences are positive. This implies that R_p is closed under operations $+$ and $\cdot$ in R and completes the proof of the theorem. ▮

One can show that, with the ordering induced from R_p, R is Archimedean and Dedekind complete. The proof of completeness is much more difficult than the one we were able to give in Appendix 2, the reason being that when one defines real numbers in terms of cuts, completeness is practically built into the definition. On the other hand, the construction we have given here, using Cauchy sequences, leads to much simpler definitions and proofs of the algebraic properties of the reals than were possible in Appendix 2.

Glossary of Symbols

Throughout the text we have used certain letters to denote certain subsets of the set of real numbers. Thus

I denotes the set of integers,
I^+ denotes the set of positive integers,
Q denotes the set of rationals
Q^+ denotes the set of positive rationals,
R denotes the set of real numbers, and
R^+ denotes the set of positive real numbers.

Listed below are other special symbols we have used, arranged in their order of occurrence. Beside each symbol there is a brief explanation of its meaning and a page reference.

Chapter 0

$\Rightarrow$	implies	*page* 6
$\Leftrightarrow$	if and only if	6

Chapter 1

$\in$	is an element of	11
$\{x \in A \mid P(x)\}$	set-building notation	11
$\subseteq$	is a subset of	12
$\subset$	is a proper subset of	12
$\cup$	union	14
$\cap$	intersection	14
$-$	complement	14
$\times$	cartesian product	14
(a, b)	ordered pair	14
$\varnothing$	empty set	14

Chapter 2

$f: A \to B$	f is a function from A into B	25
1-1	one to one	25
$\mathscr{P}(X)$	the collection of all subsets of X	25
f^{-1}	the inverse of f	27

$f \circ g$	composition of functions	28
$\{x_i\}$	sequence	30

Chapter 3

xRy	x is R-related to y	39
$\triangle$	the diagonal	39
$=, \equiv$	equivalence relations	42
$[x]$	equivalence class of x	42
$<, \leq$	partial orderings	45

Chapter 4

$\preceq, \approx$	cardinality relations between sets	64
$\aleph_0$	aleph	65

Chapter 5

$+, \cdot$	addition and multiplication of real numbers	68
(a, b)	open interval	75
$x_n \to x$	the sequence $\{x_n\}$ converges to x	77
l.u.b.	least upper bound	71
sup.	supremum	71
g.l.b.	greatest lower bound	71
inf	infimum	71

Chapter 6

$\sum$	the sigma notation for sums	102
$\sum ar^{i-1}$	geometric series	105
$a_0 \cdot a_1 a_2 \cdots$	decimal expansion	114

Chapter 7

$P_n(x; y)$	the Taylor polynomial	122
$R_n(x_1; x_0)$	remainder in Taylor's formula	128

Index